Culture and Development in a Globalizing World

T0173952

What difference does it make to add "culture" into development thinking and projects on the ground? Culture has gone from being a "background" factor in development thinking to becoming a new buzzword, seen to be central in the dynamics associated with development processes. Yet the evaluation of the practical and theoretical implications of this cultural shift in development has often been abstract.

By contrast, this volume offers a grounded engagement with culture as it enters into development paradigms, institutions, and local dynamics. With case studies ranging from Africa through to Andean Latin America, the chapters provide a detailed empirical discussion of the possibilities of, and limits to, "adding culture" into development.

Key scholars have combined broad theoretical discussions on the neo-liberal context for development's cultural turn and the concept of social capital with thorough, critical, and original evaluations of specific development processes and projects. The chapters thus bring the culture and development debate up-to-date by using the latest theoretical approaches to socioeconomic change to critically evaluate current initiatives.

Sarah A. Radcliffe lectures in the Department of Geography at the University of Cambridge.

Culture and Development in a Globalizing World

Geographies, actors, and paradigms

Edited by Sarah A. Radcliffe

Routledge
Taylor & Francis Group
LONDON AND NEW YORK

First published 2006
by Routledge

Published 2017 by Routledge
2 Park Square, Milton Park, Abingdon, Oxon OX14 4RN
711 Third Avenue, New York, NY 10017, USA

Routledge is an imprint of the Taylor & Francis Group, an informa business

Typeset in Times by
Florence Production Ltd, Stoodleigh, Devon

British Library Cataloguing in Publication Data
A catalogue record for this book is available from the British Library

Library of Congress Cataloging in Publication Data
Culture and development in a globalizing world: geographies, actors, and paradigms/
edited by Sarah A. Radcliffe.
p. cm.
Includes bibliographical references and index.
1. Community development.   2. Economic development – Social aspects.
3. Culture.   4. Culture and globalization.   5. Social capital (Sociology)
I. Radcliffe, Sarah A.   II. Title: Culture and development in a globalizing world.
HN49.C6C85 2006
306.3'09172'4–dc22
2005015958

ISBN13: 9–78–0–415–34876–8 (hbk)
ISBN13: 9–78–0–415–34877–5 (pbk)

For my son Ben, my sister Jenny, and my mother Nancy
– across the generations

Contents

List of illustrations   *ix*

Contributors   *x*

Acknowledgements   *xiii*

1   Culture in development thinking: geographies, actors,
and paradigms   1
Sarah A. Radcliffe

2   Culture, development, and global neo-liberalism   30
Michael Watts

3   Culture and conservation in post-conflict Africa:
changing attitudes and approaches   58
Elizabeth E. Watson

4   Indigenous groups, culturally appropriate development,
and the socio-spatial fix of Andean development   83
Sarah A. Radcliffe and Nina Laurie

5   Laboring in the transnational culture mines: the work
of Bolivian music in Japan   107
Michelle Bigenho

6   Social capital and migration – beyond ethnic economies   126
Jan Nederveen Pieterse

7 Social capital as culture? Promoting cooperative action
in Ghana   150
Gina Porter and Fergus Lyon

8 On the spatial limits of culture in high-tech regional
economic development: lessons from Salt Lake City,
Utah   170
Al James

9 Mobilizing culture for social justice and development:
South Africa's *Amazwi Abesifazane* memory cloths
program   203
Cheryl McEwan

10 Conclusions: the future of culture and development   228
Sarah A. Radcliffe

Bibliography   238

Index   272

Illustrations

Figures

3.1 Map of Ethiopia showing location of Regional States
and approximate location of Borana Zone 68
8.1 Hypothesized culture hierarchy in the region 175
9.1 An example of a memory cloth: no. 41 by Ntombi
Agnes Mbatha 208
9.2 Makhosi Khanyil, no. 798 215

Tables

6.1 Social capital and cultural difference 144
8.1 Utah's high-tech subsector in 2000 177
8.2 Basic distribution of the survey and case study samples 178
8.3 Unpacking the cultural economy of computer software
firms on Utah's Wasatch Front with regard to firms'
innovative capacities 180
8.4 Measuring the economic performance of Mormon
versus non-Mormon computer software firms on
Utah's Wasatch Front 192
8.5 High-tech cluster policy "shopping list" of necessary
institutions 197

Contributors

Michelle Bigenho, Associate Professor of Anthropology at Hampshire College, has authored *Sounding Indigenous: Authenticity in Bolivian Music Performance.* Her research interests include the politics of authenticity; authorship, property, and indigenous rights; folklorization and patrimony; sensory experience and meaning; and area studies knowledge production. She is currently researching the globalization of Andean music, taking Bolivian music in Japan as a specific focus of her work.

Al James is an Assistant Lecturer in the Department of Geography at the University of Cambridge, and a Fellow of Fitzwilliam College. He is an economic geographer with ongoing research interests in cultural economy, the geographical foundations of regional economic development, and geographies of work and workers in the new economy.

Nina Laurie is Professor of Development and the Environment in the School of Geography, Politics and Sociology, at the University of Newcastle, UK. She focuses on Latin American development with interests in gender, indigenous issues, and water politics in the Andes. She works collaboratively with colleagues at CESU, San Simón University, Bolivia. Her recent and forthcoming publications include articles in *Development and Change* (2005), *Antipode* (2005), and, together with Robert Andolina and Sarah Radcliffe, *Multiethnic Transnationalism: Indigenous Development in the Andes* (Duke University Press, forthcoming).

Fergus Lyon is a Reader at the Centre for Enterprise and Economic Development Research, Middlesex University, UK. He studied micro-enterprises and community groups in Ghana over a four-year period. He has also carried out research on similar topics in Nigeria, Pakistan, northern India, and the UK. He has a Ph.D. from the Department of Geography, University of Durham, UK and has published on issues of trust, associations, and economic development. Current research interests include trust in inter-firm relationships, entrepreneurship in social/community enterprises, cooperation among private sector providers of public services, and small business involvement in scientific research projects.

Cheryl McEwan is Senior Lecturer in Human Geography at Durham University. Her interests are in cultural, political, and development geographies, with a specific interest in postcolonial and feminist theories. Recent research projects have examined issues of gender, citizenship, culture, and "empowerment" in post-apartheid South Africa. She is author of *Gender, Geography and Empire* (Aldershot: Ashgate, 2000) and co-editor (with Alison Blunt) of *Postcolonial Geographies* (London: Continuum, 2002).

Professor of Sociology at University of Illinois Urbana-Champaign, **Jan Nederveen Pieterse** specializes in transnational sociology with research interests in globalization, development studies, and intercultural studies. He taught in the Netherlands, Ghana, and as visiting professor in Japan, Indonesia, Pakistan, Sri Lanka, Thailand, and Germany. Fellow of the World Academy of Art and Science, his recent books are *Globalization or Empire?* (Routledge, 2004), *Globalization and Culture: Global Mélange* (Rowman & Littlefield, 2004), and *Development Theory: Deconstructions/ Reconstructions* (Sage, 2001). Website: https:// netfiles.uiuc.edu/jnp/www/

Gina Porter is Senior Research Fellow in the Department of Anthropology, University of Durham. Her research interest in West Africa spans 30 years. Recent and current research includes work on market institutions, rural mobility, and personal networks (in NGO–state relations and among young refugees).

Based in the Department of Geography, University of Cambridge and New Hall, Cambridge, **Sarah A. Radcliffe** has research interests in social difference and development, especially in relation to Ecuador and Peru, and in social and spatial theory.

Her publications include *Multiethnic Transnationalism: Indigenous Development in the Andes* (co-author, Duke University Press, forthcoming), *Re-making the Nation: Place, Identity and Politics in Latin America* (Routledge, 1996), and *Viva! Women and Popular Protest in Latin America* (co-edited, Routledge, 1993).

Elizabeth E. Watson is a Lecturer in the Department of Geography at the University of Cambridge. She conducts research into indigenous knowledge, development and the politics of identity in Sub-Saharan Africa, specializing in Ethiopia. Recent publications include: "Examining the Potential of Indigenous Institutions for Development: A perspective from Borana, Ethiopia," *Development and Change* (2003), and "Making a Living in the Post-socialist Periphery: Struggles Between Farmers and Traders in Konso, Ethiopia," *Africa* (2006).

Michael Watts is Class of 1963 Professor and Director of African Studies at the University of California, Berkeley where he has taught for 25 years. He is currently working on a book on the political economy of oil in West Africa.

Acknowledgements

This book grew out of a set of papers presented at the 2003 American Association of Geographers annual conference, held in New Orleans, around the theme of critical approaches to culture and development. As the theme resonated with a number of development geographers' work and permitted us to debate aspects of development and thinking that were implicit in our current research, the idea of making them into a more formal project seemed to be worth pursuing.

This book project could not have been completed without a big effort from a number of people. First my contributors are all due a heartfelt "thank you" for their initial enthusiasm for the project of turning a series of conference papers, and budding ideas, into chapters; for their patience with my constant reminders about deadlines; and their polite silence about my failures to meet my own deadlines. Without them, this project would not have been as fun or as rewarding. While I appreciate all of the contributors, it is appropriate to highlight the contributions of two colleagues without whom this book would have been a distinctly different enterprise. Cheryl McEwan gave thoughtful discussant's comments on the 2003 AAG panel on culture and development, and her willingness to then write a chapter for this volume was very welcome. Liz Watson has been a great colleague during the preparation of this book, being enthusiastic about writing (and rewriting) despite many other calls on her time, and for taking time out for coffee to discuss it.

Various parts of this book were presented at conferences and seminars in the UK and the US, with invaluable input from colleagues at the "Gaining Ground: social, cultural and political processes of Latin America's indigenous people" conference at the University of Liverpool, and the Department of Geography, Dartmouth College. Discussions and email exchanges about the themes developed here have had a noteworthy impact, and thanks are due to Bill Adams, Robert Andolina, Gerry Kearns, Nina Laurie, Heidi Scott, Janice Stargardt, Andres Vallejo, and Bhaskar Vira.

Jan Nederveen Pieterse and I would like to thank Sage for permission to reprint his chapter as a slightly amended version of an article that originally appeared in *Ethnicities*, 3(1), 2003: 29–58. I would also like to thank Pion Limited, London, for permission to present in the conclusion here an adapted excerpt from a paper that originally appeared in *Environment and Planning D: Society and Space* 24(2), 2006. I would also like to thank Ian Agnew and Phillip Stickler at the cartographic room, in the Department of Geography, University of Cambridge, for their work preparing slides and figures for this book.

On a final note, I would like to thank Zoe Kruze and Andrew Mould for their patience and encouragement – in turns! – concerning this book. And my family, especially Guy, has put up with my constant discussion of this topic, and been patient (mostly) about the time spent on it – so a loving "thank you" for all your support.

Sarah A. Radcliffe
Cambridge
May 2005

1 Culture in development thinking: geographies, actors, and paradigms

Sarah A. Radcliffe

Why culture and development? The context

As I was researching Ecuadorian development projects and debates recently, references to culture kept coming up – in my conversations with Quechua indigenous representatives, in roundtable discussions about national policy, and in the corridors of Washington DC based multilateral development agencies. Not only were differently positioned actors and institutions talking about culture; people and policy were additionally drawing on specific examples of culture in action to illustrate their points. Indigenous leaders and international donors pointed out how in Bolivia's Andean highlands, "traditional" forms of decision-making and administration were giving unionist structures a run for their money. In the midst of Ecuador's economic and political crisis, Indian traders from the Otavalo area continued to export their distinctive textiles worldwide, giving anti-poverty policy-makers food for thought.

Culture has always been in development thinking and practice, but *how* it is conceptualized and *when* and *where* put in to operation reflect complex historical and geographical patterns of institutional, social, and political action. As the chapters here show, culture has recently acquired a new visibility and salience in development thinking and practice. Whereas in the past cultural norms and assumptions might have informed powerful development actors in their interaction with beneficiaries, culture is now being discovered *among* those very beneficiaries. Development practitioners and development thinkers alike are puzzling over the implications of culture for the participation of beneficiaries, for the success of projects and how culture contributes to non-economic goals of development. This volume examines cultures and the puzzles they throw up for development thinking and practice, by analyzing the "why, how, when and where" questions of culture and development. By starting from specific historical, social, and geographical locations, the chapters illustrate what happens when culture is taken seriously in grappling with development practice "on the ground."

Despite the breadth of the development field, there is no doubt that culture has arrived in development. Development thinking in the past decade has experienced a cultural turn (Chua, Bhavnani, and Foran 2000; Schech and Haggis 2000; Clague and Grossbard-Shechtman 2001), in what Kliksberg presciently termed "the new development debate" (Kliksberg 1999: 84). The emergence of culture at the heart of mainstream development debates has been a core feature of development since the late 1990s (Worsley 1999: 41). Major international initiatives such as in the United Nations Decade for Cultural Development from 1988 to 1997 have placed culture and development together, while multilateral development agencies have began to talk about the need for "culturally appropriate development" (Davis 1999: 28; UNDP 2004). Among a broad group of development practitioners from applied anthropologists through to World Bank economists, a general agreement exists that "culture in its broadest sense needs to be brought into the development paradigm" (Davis 1999: 25). Bringing culture into development, however, requires a rethinking of development's objectives and its treatment of the complex concept of culture.

The cultural values that underlie the global context for development thinking and interventions are by now increasingly widely recognized and analyzed (Escobar 1995; Schech and Haggis 2000). Studies have demonstrated how postcolonial legacies of cultural interaction and the contests over development's meanings and practices have long been questions of culture. Yet as traced in this introduction and the following chapter, a recent paradigm shift has occurred in development's approach to culture as cultural difference is now treated explicitly as a significant variable in the success of development interventions (Rao and Walton 2004). In order to understand the reasons for this new development debate, however, we must first look at how development itself has been conceptualized. In the mid-twentieth century, development was equated with poor countries' economic growth and modernization that were expected to replicate Western experience. To summarize a complex history (see Schech and Haggis 2000; Watts, this volume), development thinking was increasingly challenged by Marxist, feminist, and postcolonial writers and activists, and began to reconsider its own specific institutional, historical, and cultural location. The approaches of these critical writers demonstrated how development included not only the specific interventions – projects, programs, loans, and aid flows – usually included in definitions of development, but that it was additionally embedded in the cultural economy of Western capitalist political economies and the cultural histories of European colonialism.

In current understandings, development includes the reworkings of relations of production and reproduction, and of sociocultural meanings, resulting from planned interventions and from uneven political economies. Development comprises "an uneven motion of capital finding, producing and reproducing places and people in particular and differentiated relation to peculiar strategies of accumulation . . . [Its] signal form in the second half of the twentieth century demarcated a specific relationship between the global north and south or between the 'first' and 'third worlds'" (Katz 2004: ix). As it encompasses both intentional practice and broad political economic processes (Cowen and Shenton 1995: 28), development's double-sided essence has to be kept in creative tension in any discussion of development's constantly shifting horizon of global and local change (Hart 2001).

The context for taking culture seriously in development arises from a number of cognate issues, processes, and debates. In order to explain development's "cultural turn," commentators identify five main reasons for the recent prominence of culture as a key concept in development thinking. These reasons include the failure of previous development paradigms; perceptions of globalization's threat to cultural diversity; activism around social difference (gender, ethnicity, anti-racism); the development success stories in East Asia; and the need for social cohesion. These views are not found together, and even one reason for taking culture seriously covers a number of different political, analytical, or ideological perspectives.

One of the key prompts for a rethinking of development's relationship with its cultural field was the widespread disillusionment with development among practitioners, thinkers, and grassroots actors from the 1980s. While the impasse in development thinking was argued by sociologists, anthropologists, and geographers to be due to the inability of development thinking to overcome its economism and teleological frameworks, the practical failure of projects on the ground to deliver satisfactorily was a key component (Nederveen Pieterse 2001a: chapter 3). During the 1980s in many parts of the majority world, development indicators were reversed due to the combined effects of debt burden, falling productivity and job availability, and loss of development directions. In Ecuador for example, rural credit and various development programs of the Inter-American Development Bank and UNDP were suspended or terminated due to conflict and poor project performance (Griffiths 2000: 33). The following chapters give other examples of development rethinking following negative experiences with development work.

Prominent in discussions about culture in development are concerns about the potentially homogenizing cultural effects of globalization. In the words of the United Nations report on culture and development, a "danger looms of a uniform global culture" (United Nations Report 1998: 22). Voices from the global South also raise the specter of loss of cultural diversity. In Africa, the erosion of cultural heritage under the experience of development is argued to "precipitate development crises" causing alienation and disorientation for ordinary Africans (Yakubu 2002: 8–9). Similarly, Prah argues that "the brooding presence of Western culture [in Africa] is singularly blighting and fossilizing indigenous cultures" (Prah 2001: 96). During its Decade on Culture and Development (1988–97), the United Nations argued that "the defense of local and regional cultures threatened by cultures with a global reach" (UN Preamble 2003) required in turn action for "preservation of the diversity of cultures" (UNESCO 2003). According to development anthropologists, the way in which globalization tends to lead to the "leveling of national and local cultures and its consequent social dislocations and economic crises" (Davis 1999: 31) needs to be addressed. In order to challenge or halt cultural homogenization, cultural policies beyond the national scale are being promoted at inter-regional and even global scales (UN Report 1998: 22). Such policies can provide "an antidote to globalization" according to the Dutch Minister for Development Cooperation (quoted in UN Report 1998: 85). Yet cultural globalization occurs in the context of global markets, complicating the agendas of democratic national policies in their attempts to shape increasingly complex individual consumption patterns (Sen 2004: 52–3). Trajectories of desire often encompass goods produced around the world, making the substitution of "national" products for "global" products problematic (Sen 2004: 52–3).

Anti-discrimination measures connected with postcolonial political struggles have also played their part in questioning the implicitly Western cultural focus and expectations that long underlay development thinking and practice. Efforts to challenge Eurocentric visions of progress and development modernity thus form part of a broad agenda to question culture's power to proclaim what is an appropriate or inappropriate culture in a development context. For example, some African commentators argue that development occurring after colonialism brings a profound malaise to the region's cultures by undermining their autonomous values (Prah 2001). As a result of these broad critiques, development policy has explicitly attempted to understand and encompass a regard for cultural diversity during the last decade (Allen 2000). Often analyses from these perspectives are informed by diverse Third World feminist movements, and by campaigns to ensure rights for populations marginalized by ethnic/racial hierarchies (such as the global indigenous rights movement). While in fact most people around the world live with the overlap and juxtaposition of multiple cultures (produced by variable combinations of local societies, nation-states, international consumer and religious cultures), anti-discrimination and anti-racist work has to grapple with the persistent hierarchies between cultures and groups in multicultural and multiethnic societies. Building bridges of communication and mutual respect creates intercultural understanding and speaks to an agenda of working across/through social differences. Hence the United Nations' "commitment to pluralism" works from the need to build policies around cultural diversity and intercultural understanding (UN Report 1998: 22), a position endorsed by applied anthropologists (Davis 1999: 28). The economic benefits of working for intercultural understanding have also been pointed up by development economists who argue that development of flexible labor markets and employment opportunities are impeded by employers' misperceptions of workers' cultural attributes, thereby causing the waste of human capital (Kliksberg 1999; Hojman 1999).

The dramatic growth and improved living standards for populations in certain East Asian countries, often termed the Asian Tigers, has generated considerable debate about what lies behind their success story. Early economic discussions attributed much weight to the region's ancestral and long-standing cultures, arguing that the Confucian tradition provided sociocultural rules that assisted accumulation. Such arguments finally put to rest the assumption, embedded in mid-twentieth century modernization models of development, that Western-style development was the only trajectory to growth (Worsley 1999: 34). Certainly, the West is no longer a privileged interlocutor in definitions of development and modernity (Nederveen Pieterse 2001a; on Japan, see Goodman 1999; Sen 2004: 48–9). Yet attributing a broad historical influence to culture was ultimately unsatisfactory as it attributed a coherent cultural web over highly diverse meanings, practices, and social relations (Ong 1999). Analysis began to focus instead on daily performances of cultural economic practices that underpin, say, Japanese business behavior, such as the "after-hours sessions in the bars and nightclubs . . . where the vital personal contacts are established and nurtured slowly" (Lohr, quoted in Granovetter 1985: 67). In other words, the Asian Tigers demonstrate *variable* organizational and social cultures through which economic transactions and values are expressed and reproduced. In Singapore for instance, a complex interaction between social Darwinism, Confucianism, and specific leadership styles contributed to its development experience (Chang 2002). By focusing on the state and the firm as locales for culture, distinctive practices and meanings have been identified in the forms of governance, workplace dynamics, and education that are now thought to contribute to economic growth. By examining culture as a mode of organization at a number of levels therefore, the selectivity and flexibility of cultural relations across spheres of production and reproduction have been highlighted without reducing culture to a "catch all" category (also James, this volume).

Another agenda behind development's cultural turn is the objective of overcoming tensions and potential conflicts between human groups. Departing from the insight that the resource of culture is held by every individual and social group regardless of their economic or political power, a number of different strands of policy have emerged. Emphasizing the non-economic facets of development, the right to freedom of cultural expression represents a development goal that depends upon security, democratic openness, and accountability

(Friedmann 1992; Sen 1999). One of the first contexts for development's explicit attention to culture was in conflict resolution, where tension and violence were attributed to a lack of cross-cultural understanding, and where the rebuilding of post-conflict societies departed from the cultural resources of local groups (Davis 1999: 37; compare Watson, this volume). In this vein, UNESCO promotes an agenda of social cohesion to attempt to overcome potential conflicts and inequalities along a number of axes of social difference (UN Report 1998).[1] Such approaches offer a constructive response to heightened global security concerns that are wrongly reduced to culturalist explanations (see below for a discussion on culturalist explanations). "Dialogue between . . . flexible, multiple and open identities and cultures should become the basis for a concord to cultures rather than a 'clash of civilizations'" (UN Report 1998: 23; Rao and Walton 2004: 10). The promotion of social cohesion and rights to cultural expression has recently been at the heart of measures to recognize the cultural capital of the poor. Whereas impoverished people may have little more than cultural identities, this cultural capital is viewed in recent policy as the launch pad for transforming their relative position in multicultural societies. However, the means of achieving a change in status is far from straightforward (UNESCO 2003), as the following chapters illustrate.

As Bjorn Hettne argues, the new emphasis on culture has far-reaching implications and constitutes a major challenge to the rethinking of development (Hettne 2001: 9). Moreover, the varied engagements of development with cultural questions outlined above give rise to *different* trajectories for theorization and policy-making. If cultural development is a response to anxieties about globalization, then policy measures to enhance cultural diversity in global world become adopted, as illustrated by various United Nations initiatives. Tackling discrimination raises different challenges for policy, not least the difficulty of transforming the terms of knowledge, power, and hegemony of Western (colonial and contemporary) cultures of modernity and development into more plural, open, and empowering mechanisms. If, by contrast, the development agenda is informed by the Asian Tigers' case, policy questions are oriented to bring about the necessary managerial, work culture, and political transformations within a capitalist political economy. Yet each distinct concern brings us back to the fact that development thinking and practice has, over the past 10 to 15 years had to rework its understandings of development while refining its conceptualization of culture.

What is culture? What is development?
Debates across disciplines

Culture has come out of the seminar room discussions and gone into the manuals for development fieldworkers and institutions. Yet the term "culture" has no universal or agreed meanings, and indeed varies in its significance and theoretical background across anthropology, development studies, geography, and cognate disciplines. Drawing on recent debates in a number of theoretical and substantive fields, this section outlines a broad-based characterization of culture. We explore in detail the ways in which culture and development have been defined and used by key disciplines in order to work towards an interdisciplinary understanding of how culture and development can be conceptualized in relation to one another. Anthropology and development studies – focused on issues of culture and development respectively – start off this discussion, before turning to critical approaches (including feminist and post-development work) and then to economics and economic geography. The aim of this section is to work towards multidisciplinary definitions of development and culture that are mutually comprehensible.

Anthropology

Although central to the discipline of anthropology, culture remains a highly ambiguous concept and one that arouses much controversy (Gardner and Lewis 1996; Fox and King 2002; Mitchell 2002). Whereas historically focused on small-scale societies with contiguous cultural boundaries, anthropology increasingly has to engage with a global stage in which cultural mixing, hybridity, and multiple and crosscutting flows of people, meanings, and artifacts are taken for granted (Gupta and Ferguson 1997; Buroway, Blum, and George 2000). Anthropology's own complicity in the hierarchical labeling of Other cultures has been traced back to colonial times (Abu-Lughod 1999). Under colonialism, although "Western civilization" was pronounced to be the model for non-Western societies, in practice non-Western culture was selectively regularized to facilitate colonial rule and early postcolonial politics (Worsley 1999: 35–6). Colonial models of power attributed culture with a central significance in the ordering of the world.

However, to attribute culture with the ability to determine all aspects of social life is to fall into the trap of culturalism whereby any social feature is linked causally and directly back to a broad category of "culture" (Worsley 1999: 37). Culturalist explanations tend to assume that culture is an unchanging "bundle" of beliefs and practices that are coherent and homogeneous across a large population. In recent years, culturalist explanations have gained widespread media and popular prominence; broad-brush associations between religious or historical identifications have been offered as the explanation for political attitudes and responses to modernity. Culturalist frameworks including Islamic/Christian and East/West binaries claim to differentiate between groups yet they suppress long-term historical patterns of connection and internal cultural diversity in their search for headline "explanations" (Ong 1999; Hart 2002; Berger 2003). In such binary frameworks, cultural difference is reduced to being a source of conflict or a source of functional integration (Huntington 1996; Harrison and Huntington 2000; Landes 2000), although in each case culture is accorded a unity and primacy that anthropologists abandoned long ago.

Culturalist explanations in effect ignore the key dimensions that anthropologists now place at the heart of the culture concept, namely its flexibility, its strategic deployment by different actors to make interventions, its contested content, and the interplay between material and symbolic components.[2] By taking culture for granted – as a "traditional way of life" (Werlin 2003: 337) or as a template for action – culturalist approaches ignore struggles over cultural meaning, the state's role in shaping culture, and the coexistence of different cultural registers. In practice, culture is used flexibly, each variation in meaning and each social relationship passed – or not – through time or over space, mobilized or forgotten in turn (Fox and King 2002). Hence in order to retain a robust definition of culture, its limits need to be recognized. While culture *does* include a broad range of components such as the discursive understanding of contested meanings (e.g. elite versus popular culture), material culture (which embodies socioeconomic organization, meanings, and inequalities), structures of feeling, and forms of social organization (such as kinship, religion), it does *not* explain or control everything in social life.

One key limit of culture's power came with the recognition that in the past the concept was too often equated with – or treated as a coded word for – racial difference. From the colonial period into the 1920s, the term culture became to be interchangeable in the West with the new notion of ethnic group, and both concepts substituted for race (Visweswaran 1998: 76). In such a way, today culture can operate as an alternative term to race. However, race/ethnicity and culture cannot be collapsed down into each other, despite the concepts' entangled history in Western thinking and policy. Instead, the separate yet interconnected realms of culture and race need to be acknowledged, and in each case the historical and power-infused relationships that underpin a society's "culture" and its racial relations have to be acknowledged (Trouillot 2002: 41). Racial categories are not biologically given; rather, "the social process by which racial categories are created, inhabited, transformed and destroyed" comprises a society's racial formation (Omi and Winant 1994: 56). Racial formations influence how resources are distributed among groups, yet are contested in and through different racial projects (projects carried out by the state, or by social movements). Thus culture in development requires an analysis that reconnects it with class, history, and racial formations. Just as political racial projects combine systems of meaning with the allocation of entitlement, so too does development (White 2002: 416). For example by reworking the meanings of racial identity and the distribution of development aid, pro-indigenous development projects in the Andes represent new racial projects that reconfigure national racial formations (Radcliffe and Laurie, this volume).

Development studies

Separating out culture from race occurred in development studies only recently (White 2002) as for much of the twentieth century, the discipline relied upon culturalist frameworks to explain underdevelopment of former colonies and countries in the global South. In the mid-twentieth century, the main development paradigm was modernization. While Walt Rostow's modernization paradigm implied that American cultures of consumerism represented the epitome of development, Talcott Parsons argued that certain habits of mind and behaviors would benefit modernization. In this way, development thinking took on board the notion that culture can be acquired, and results from a learning process. Non-Western – "traditional" – cultures were perceived as fading relics that would inevitably and unproblematically be replaced by modern forms of culture and identity as people learnt Western ways. Mainstream accounts through much of twentieth century assumed that people learn about *Western* modernity, and acquire characteristics that favor them in a capitalist, state-ordered society. Particular local cultural features were treated by modernization thinking as barriers to the expansion of western cultural attributes and development.[3] In different contexts and times, agricultural involution, over-close ties with family and kin, cognitive maps, and the culture of poverty were all perceived in modernization thinking as blocks to (Western) development (Allen 2000). Yet Gunnar Myrdal, a key development thinker through the mid-twentieth century, recognized Western approaches' lack of universality. For many years following modernization debates, development's engagement in questions of culture was denied due to its colonial legacy and then modernization's culturalist explanations (Worsley 1999: 30).

In recent years, development studies has moved away from these approaches to embrace a critical and globally informed perspective on the nature of development and its interventions. Key to this change has been the recognition that Western notions of change are discontinuous and divergent as different social, political, and economic interests are bundled up into any one policy (Nederveen Pieterse 2001a: chapter 3). At the same time, the ideas and practices of development – as immanent process of socioeconomic transformation and as specific interventions – are re-embedded in locally situated practices (Arce and Long 2000: 1). This creates a highly nuanced picture of development as it works its way over variable situations and encounters variable combinations of actors and their interests. Moreover, strategic actions to counter the dominant development trends exist in every society, comprising what Arce and Long call "counter-tendencies." These counter-tendencies rework notions of modernity and development, departing not just from "traditional" culture but through recombining old and new modes of working and meaning (Arce and Long 2000: 21). Focusing on the middle ground of interactions between actors and institutions, development studies is concerned with how interpretations of development and modernity, along with strategies to deal with them, are worked out across a multiplicity of interconnected sites (Arce and Long 2000: 21). In this approach, understanding development engages what Nederveen Pieterse terms a "critical globalism" that involves "theorizing the entire field of forces in a way that takes into account not just market forces but also interstate relations, international agencies and civil society in its domestic and transnational manifestations" (Nederveen Pieterse 2001a: 46). By re-engaging with the social field in which development operates, development studies has offered a persuasive account of the actors, institutions, and multiple sites through which development is thought, operationalized, and experienced.

Critical accounts

Critical accounts of culture and development challenge the ways in which mainstream categories and relations underpin relations of inequality and exclusion. Using a post-structuralist critique of discourse and articulating the possibility of more equal societies, critical accounts during the course of the 1990s provided key insights into the relationship between culture and development. Post-structuralist analysis of the discourses of development has documented the numerous cultural assumptions that underpin development thinking (Crush 1995; Schech and Haggis 2000). They have had an impact across a number of disciplines including anthropology, development studies, geography, and sociology. Key to these critical and post-structuralist approaches are Third World feminist accounts and post-development writings respectively.

Third World and postcolonial feminisms were among the first critics of the underlying Western normative frameworks that underpinned much development in the twentieth century. Critiquing accounts of Third World women as homogeneous passive victims of patriarchal local cultures whose liberation would occur through development (Mohanty, Russo and Torres 1991), Third World feminists reinserted women's agency, diversity, and cultural distinctiveness into development thinking (Schech and Haggis 2000, Chapter 4). Many women in the global South have less secure access to the uneven opportunities of development than their male counterparts, due to the way that hegemonies of political economy, law, kin relations, religion, and cultural tradition come together to shape local social relations and structures of feelings (Grewal and Kaplan 1994). While gender and development agendas to introduce feminist concerns into development do address issues of women's inequality and represent an improvement on earlier policy approaches (Kabeer 1994), they remain implicitly modernizing and Western (Bulbeck 1998). New practices are required in order to recognize and encourage the creation of diverse path-dependent and non-linear opportunities for gender equality that avoid Western norms (Jolly 2002). Treating culture as lived experiences, structures of feeling and the relationship between production and reproduction, postcolonial feminist approaches resonate closely with new anthropological definitions of culture, yet additionally stress the ongoing significance of gender difference in determining individual development outcomes (Chua, Bhavnani, and Foran 2000). Development interventions can challenge, rework, or reinforce the cultures of gender within which women and men live (Jolly 2002). Reworking connections between culture, women and development entails a focus on women as agents of change in diverse spheres of life and recognition of women's class, ethnic-racial and social diversity (Chua, Bhavnani, and Foran 2000; Bhavnani, Foran, and Kurian 2003).[4]

Culture and power are placed at heart of development debates by post-development writers (Worsley 1999: 39). Attacking the Western enlightenment binary forms of thinking (First World/Third World; rich/poor; white/black) which entered into development thinking and practice, post-development brings a useful critique to the implicit cultural assumptions that underpinned development thinking through from colonial times to the present. In searching for alternatives to the hierarchies in development paradigms and practice, post-development writers highlight the role of culturally distinctive actors in contributing to a "post-Western" development model. Post-development tends to focus on the analysis of social movements and grassroots actors, who are often viewed as culturally distinctive to national culture (Escobar 1995; Apffel-Marglin 1998). Yet in celebrating local cultures as alternatives to "standard" development, post-development writers risk overemphasizing the difference between the West and the "Rest," while ducking the question of how to engage with useful definitions of culture and its complex local connections with race, economics, and politics (see Lehmann 1997). Postcolonial approaches that analyze the complexities of recent political economies and attendant cultural meanings are better placed to deconstruct changing development paradigms.

Economic geography and development economics

Although the Washington consensus on neo-liberal policy frameworks became hegemonic globally during the mid-1980s, its ascendance was matched by the increasing recognition of how varieties of capitalism exist across the world, rooted in different historico-geographies of cultural practice and institutional settings. At the same time, the taken-for-granted idea that the economy and culture referred to separate spheres of activity was increasingly questioned (Trouillot 2002: 55; Amin and Thrift 2004: xii). "From the standpoint of economics, cultural activities have been regarded as a secondary field foreign to the central line that economic growth should follow" (Kliksberg 1999: 96). Previous models of the economy had treated individual actors as either free of cultural reference or totally subordinate to social norms. Yet, it was increasingly recognized, "actors do not behave or decide as atoms outside a social context, nor do they adhere slavishly to a script written for them by the particular intersection of social categories that they happen to occupy" (Granovetter 1985: 58). Drawing on the notion of culture as changing and embedded in specific institutional and social relations, Britain's "complacent traditionalism," for example, was blamed for its poor economic performance (Hickox 1999: 137).

The ongoing dynamic influence of institutional and sociocultural context became increasingly recognized, as the fractured and unpredictable turns taken by cultural change, social relations, and meanings became visible (Clague and Grossbard-Shechtman 2001; Sen 2004). Understanding the economy's social and institutional embeddedness opened up new questions and tentative answers about trust and its downside of corruption, the specific social practices (gossip, socializing, networking) through which business got done and labor was recruited (Granovetter 1985). Uneven economic development was no longer a question of management of profit goals in financial terms; it was attributed to the dynamism of regional cultures where face-to-face contacts, trust, and stable social networks underpinned ease of access to business information, labor, and credit (Putnam 1993). For example, whereas Silicon Valley, California, is held up as an icon of the synergies of culture and economy, the "Silicon wannabes" – that is, places wanting to repeat Silicon Valley's success – are seen not to combine the "right" cultural features (James 2006: 7).

Recognition of cultural embeddedness of economies pinpoints the institutional and regional dimensions of economic growth processes, yet often underestimates the complexity of spatial connections and the non-institutional sociocultural processes that contribute to economic behavior (Barnes 2002). By rejecting an individualistic *homo economicus* actor, Granovetter's approach underplayed the role of specific actors – firms, financiers, nation-states, unions, individuals among others – in determining the interplay between pursuit of prosperity and other goals. Regional dynamics thus became a function of abstract processes "rather than . . . the deliberative human agents, bureaus and groups responsible for those processes, along with the spatial structures that shape their actions" (James 2006: 5). Redressing the lack of focus on multiple institutions and actors, attention has been paid to cultural entrepreneurs, who – it is argued – mobilize and support productive trends and the emergence of new consumption patterns. Within this policy model, the cultural entrepreneur plays a key role creating alliances between "the new entrepreneurial middle class, potential winners, [and] opinion formers" to create dynamic economies (Hojman 1999: 180, 177). Yet cultural entrepreneurship occurs within wider socioeconomic transformations that are moving towards fragmented working lives and the loss of welfare systems. Precarious employment conditions, work fragmentation, multiskilling, and multitasking are the experiences of many workers in the global post-Fordist economy that relies upon just-in-time production and adaptable workforces to maximize the flexible accumulation of capital (Harvey 1989). For example, in the European Union, artists-turned-cultural workers are expected to be entrepreneurial (raising finance, seeking out new markets) all the while maintaining high standards of cultural production (Ellmeier 2003). Embedded within similar processes, Bolivian musicians touring Japan to give school concerts are not mere entrepreneurs riding the wave of global economic opportunity, but work to time and to others' standards in order to make a living (Bigenho, this volume).

Furthering these accounts, economists today view the two dimensions of culture and economics as equally important and as intrinsically and simultaneously co-influencing. Just as economic transactions and dynamics increasingly came into focus, so too the ways in which culture and economy were to be brought together conceptually gained attention. To overcome the limitations of the notion of cultural embeddedness,[5] culture and economy can be conceptualized as coequal and coterminous elements as they "exist in dialectical relation, based upon their perpetual and simultaneous (re)construction by human agents whose economic motives and logics derive from their own socio-cultural identities" (James 2006: 2). The pursuit of accumulation is simultaneously the pursuit of many goals, symbolic and pleasurable, responding to "imaginaries of desire and desirability" as much as profit (Amin and Thrift 2004: xiv–xv). If culture is appreciated as economic practice and the economy as cultural practice (Amin and Thrift 2004: xviii), then development may be considered as projects to set up processes of accumulation that exist within multiple projects for pursuing other (social, cultural, political, symbolic, pleasurable) goals. By adding in an anthropological understanding of how institutions and hierarchical social relations frame individual actors' sociocultural identities, the concept of cultural economy can be used to frame the terrain within which development engages with culture.

*

In summary, culture comprises the material products, patterns of social relations, and structures of feelings produced by multiple actors, who are differentially positioned in power relations, political economies, and social reproduction. Viewing culture in this way recognizes the contested nature of cultural meanings, artifacts, and social relations that are coproduced by diverse actors in their ongoing daily and generational interactions. The spatial and social limits of culture are thus dynamic, reflecting the ways in which social interactions are rarely bounded, while the fields within which meanings and social relations are produced go beyond the local or indeed national arena. We can thus view "culture as a terrain in which politics, culture and the economic form an inseparable dynamic" (Lowe and Lloyd 1997: 1). Economies are not separate from the sociocultural field in which they occur, as the transactions, values, and institutions often bundled up into the concept "economy" are intrinsically connected – and owe their meanings, reproduction, and contestation – to culture. The cultural economy of development interventions thus encompasses the geographically variable priorities and social transactions by which social life is reproduced. Culture in this sense is intrinsic to the various projects of development. Whether in the form of uneven political economies or planned interventions, culture as a terrain of material production, structures of feelings and patterned social relations imbues development with a wide repertoire of resonance and a social backdrop which analysts have only recently begun to recognize. In a world living through the after-effects of colonialism and the diverse racial projects enacted by nation-states, culture exists within complex historically and geographically variable racial formations. Similarly, development's culture is embedded within a postcolonial global racial formation and interacts with national cultural economies of development. By locating culture in this broad terrain, development actors and paradigms can be analyzed fruitfully as interacting to produce cultural meanings, social relations, and material cultures across a number of spaces. Bringing culture into issues around development, we come to a working understanding of culture in the field of development that underpins the subsequent chapters.

A new development paradigm? Culture and development

> Development in the 21st century will be cultural or nothing at all.
> (UNESCO 1997, quoted in Carranza 2002: 36)

Culture has become a significant – and widely acknowledged – facet of development thinking in recent years, as it focuses attention on diversity and complexity and deals with issues of cohesion (Worsley 1999: 30; Kliksberg 1999). Although the conceptual starting points for this acknowledgement of culture's significance are highly diverse (see above), and in many cases theoretically incompatible, it demonstrates that culture and development are now widely perceived as dialectically related. While it is premature to claim there is a new and coherent paradigm around this pair of concepts, it is nevertheless useful to mark the terrain shared by a number of different disciplines and actors and which takes the terms jointly into the heart of mainstream development debates (Worsley 1999: 41). In this context, "culture and development" refers to the fact that culture is not primordial but is reworked and reproduced around and through development, just as development (as political economy and as planned intervention) is embedded in "imaginaries of desirability," material culture, and social relations. Placing development in a terrain beyond the unidimensional measure of GDP or the multidimensional measure of capabilities (Sen 2004), the paradigm of culture and development captures the recognition that culture is intrinsic to development as the priorities, goals, and outcomes of social life/development arise at the interface between development and culture. Culture and development, what Nederveen Pieterse terms C&D (2001a), thus goes beyond the question of whether culture is taken seriously in development (that is, whether it has been "added in"), to examine *where*, *when*, and *how* (specific historico-geographical formations of) culture and development interact.

Culture and development departs from the premise that cultural values such as expectations of material culture and structures of feelings underlie the paradigms and policies of development.

However, it also recognizes that the enactment of cultural development depends on the configuration of local, national, and global power relations that shape projects or interventions (Schech and Haggis 2000; Power 2003). (The term cultural development is used here in the spirit of the concept of cultural economy, discussed above.) For a number of distinct actors, culture increasingly represents a key factor in development outcomes, increasing the meaningfulness of interventions for project beneficiaries and the social sustainability of projects for their administrators and donors. In East Africa, the ways in culturally specific local institutions play a role in natural resource management works from the culture and development premise that particular local configurations of meaning, structure, and reproduction are linked to potential improvements in development outcomes (Watson 2003). In the Ecuadorian Amazon region, a regional NGO attempts to create a ceramics industry and tourist attractions, arguing that culturally appropriate models are more sustainable and promote local participation (Wilson 2003). In the face of destructive cultural and economic reforms, culture appears as the "missing link" in African development (Prah 2001). Launching the Agenda 21 of Culture, mayors from over 400 cities and towns around the world met in Barcelona in May 2004 to protect public spaces as spaces for multicultural encounters between locals and non-local groups (Cia and Martí 2004). These examples highlight the multifarious ways that grassroots and development actors and institutions view culture and the diverse means at their disposal to "use" culture for development ends.

Culture and development also engages diverse actors at scales ranging from the local through to the global and international (see section below). The United Nations Decade for Cultural Development, which ran from 1988 to 1997, raised international awareness of the issues, initiating international debate around cultural priorities to be brought into policy. The Decade stressed the importance of acknowledging the cultural dimensions of development, enhancing cultural identities, broadening participation in cultural life, and promoting international cultural cooperation (UNESCO 2003). In January 1998, 20 major development organizations created a network on "Cultural heritage and Development" to exchange strategies on building from cultural heritages. At the end of the United Nations Decade, conferences and symposia provided a forum in which to provide guidelines, by integrating cultural policies into human development strategies at international and national levels.

Development thinking now "contain[s] a more pro-active vision of incorporating a cultural dimension into the development process itself" (Davis 1999: 27). Such a "cultural lens" places culture at the center of development, viewing it as neither inherently damaging nor beneficial (Rao and Walton 2004). A critical engagement with culture and development paradigms adds greatly to our understandings of development institutions, processes, and practices without distracting our attention from the materialities of poverty, inequality, power relations, resource distribution, and the vagaries of the global economy (McEwan 2003: 2). Nevertheless, as argued by development economist Amartya Sen, the "import of culture cannot be instantly translated into ready made theories of cultural causation" (Sen 2004: 55). As the chapters in this volume illustrate, taking culture seriously as a factor in development does not automatically lead to a policy template to be applied across the diverse and constantly cultural contexts of the impoverished world.

Geographies and actors in culture and development

Culture has always been central to our understanding of the development processes and their impacts across the globe. Yet many development interventions have failed because of a lack of understanding of local cultures. Nation-states have intervened systematically in postcolonial countries to shape the racial formations and cultures in their territories, often tying national objectives to development targets and imaginaries (Gupta 1998). For these reasons, a detailed understanding of the ways in which diverse *actors* bring culture into development and *where* they do so, is required (in addition to how and when). If actors and institutions remain invisible in accounts of development, responsibility for change cannot be attributed to actors with agency, while an aspatial account of cultural development's terrain limits our understanding of innovative policies and practices which can be made more widely available.

Development and culture are defined crucially by the openness and contested nature of the social terrains over which they operate. Yet the specific configuration of development and culture in operation occurs together in specific locales and spaces. Development and culture are hence intrinsically constructed, reconstructed, and deconstructed at multiple scales simultaneously. As the following chapters show in detail through diverse case studies, development is adapted in national-regional contexts, varying greatly with political economies, histories, and geographies (Crush 1995). Similarly, global notions of what development should comprise are given local and regional inflections and interpretations as they become "indigenized" in different localities (Nederveen Pieterse 1998: 365; Gupta 1998). In some cases, development models are highly specific to regional contexts within which they are elaborated and applied via policy. Urban inequalities and massive rural-urban migration in Latin America for example, became the context for notions of the "culture of poverty" in the mid-twentieth century (Lewis 1959), whereas the rural itinerant herding societies in sub-Saharan Africa were brought into policy concerns via the concept of the "cattle complex" (Allen 2000).

As connections between localities and regions in a globalizing world reflect power relations and deliberate differentiation between scales, the chapters show how scale, space, and connections shape the debates and practice of culture and development. The challenge is to retain a multi-scalar framework that identifies the relevant scales – and their associated actors – involved in any one example of cultural development on the ground. Economists and geographers have analyzed the locally bounded social relations and cultures of communication and trust among business people, who belong to the same country club or exchange information over a drink in the pub (Granovetter 1985). Such cultural economic practices are pictured primarily at the local level; they exist within or alongside the working cultures of firms or a specific economic sector. Anthropology has long contributed to our understanding of social relations and cultural meanings of localized societies in the global South, highlighting the ways in which "local" beneficiaries of development are positioned in complex hierarchies of class, gender, generation, and race-ethnicity (Ferguson 1994).

At a different scale, development agencies, non-governmental organizations, nation-states, and firms each have their own (often unique) institutional working cultures, cultures that profoundly shape the ways in which development priorities can be expressed, gain support, and be implemented inside and outside the organization. Organizational cultures, reflecting the societies in which they are embedded, are characterized by ongoing attempts to influence resource distribution, priorities, and agendas (Lewis et al. 2003). Although development agencies have long been staffed by economists (and in the UK at least, former colonial officials [Kothari and Minogue 2002]), the staffing – and hence the agendas and debates – in agencies is beginning to change. Multilateral development agencies' recent enthusiasm to employ anthropologists in order to further agency work on cultural development raises new questions about the role of networks of anthropologists in institutions dominated by economists (Gardner and Lewis 1996). By the late 1990s, the World Bank social development office employed over 100 social scientists, most of them anthropologists and most of them from the global South, while in the UK development agency the number of anthropologists went up thirtyfold (Davis 1999; Eyben 2000).

At a broader scale again, multilateral development agencies play a significant role in shaping the nature of cultural development paradigms and policies (Arizpe 2004). The World Bank under President James Wolfensohn from the mid-1990s worked on the principle that there was a "need for greater sensitivity to and investment in national and local culture in the Bank's new development agenda" (Davis 1999: 26). Establishing a social development taskforce in 1996, the World Bank began to consider how effective economic development was rooted in "institutional decision-making, civil society participation, gender, ethnicity, and other forms of diversity" (Davis 1999: 25; Alkire 2004). While this agenda prioritized economic wealth above other "imaginaries of desirability," it led to institutional reform in the Bank and the appointments of social anthropologists. Fierce debates raged in the Bank as a result about how culture – framed by the concept of social capital – could be used to inform policies and practice (Bebbington, Guggenheim, Olson, and Woolcock 2004). The Bank also began to think in terms of cultural learning, prioritizing loans that would encourage flexibility and new sources of knowledge and economic activity. Treating learning as a means to transmit culture in non-linear and institutionally specific ways reinscribed culture-as-a-discrete-unit models, yet framed Bank loans to Peru and Bolivia in the mid-1990s. Multilateral development agency treatment of culture draws on understandings that inserting culture into development will increase the participation of civil society actors, guarantee social cohesion and regularize decision-making (Rao and Walton 2004). Yet while the non-bounded and deterritorialized cultures and meanings that circulate in a global field have been increasingly recognized (Gupta and Ferguson 1997; Buroway, Blum, and George 2000), their role in shaping culture and development relations remains little studied (but see Arce and Long 2000).

Culture produces, and in turn is produced by, interactions at different levels which complicates the ways in which we might attribute

culture a role in development thinking and policy. Whereas policy
tends to identify a specific target area for development interventions,
the factor of culture in this designated area might not correspond
to those policy geographies, with unforeseen consequences for the
outcome of interventions. Moreover, policy knowledges are often
not attuned to the subtle interactions between scales and differently
positioned actors in the development beneficiary region. In order to
unpack these mutually influential scales and patterns of culture, the
chapters here analyze the ways in which constructions of culture
are scalar and intersect with policy agendas in complex and
unpredictable ways. Al James (this volume) analyzes the interactions
between corporate cultures, regional *industrial* cultures, and broader
regional cultures to understand the impact of Mormon culture
on high-tech firms in Utah. Sarah Radcliffe and Nina Laurie
(this volume) examine how regional cultures interact with new
development institutions' concepts of cultures and local patterns
of ethnic society to produce development projects. Development is
always site specific and culturally distinctive yet is always
and everywhere inserted into broader cultural and geographical
groundings, which makes it ineluctably "cultural geographic" (Watts
2002: 435; also Hart 2002).

Culture and development thinking: new paradigms, recent concepts

Development thinking could not take culture seriously without
having some key concepts and frameworks for understanding how
culture and development "fit" in any given development context.
These concepts underpin and provide theoretical justification and
practical frameworks for thinking about cultural development and
practice. Three key concepts are participation and governance, as
well as social capital. During the 1990s and into the twenty-first
century, they have provided the talisman and framework for much
cultural development thinking. As such, the concepts appear in
various ways in the specific case studies discussed in the chapters
here; below, the relationship between governance, participation, and
culture are outlined, before moving on to a brief discussion of social
capital.

In the context of neo-liberal reworkings of political economy, the
state's role in ordering and managing accumulation has been

downplayed in favor of the "free hand" of the market. As a result, the political infrastructure of the state has undergone a deep restructuring, with civil society being granted a relatively larger role in organizing reproduction and production (Nelson and Wright 1995). Combined with dissatisfaction around the previous top-down models of development, in which stereotypically the state owned key resources and devised Five Year Plans to be met by industry, agricultural sectors, and resource extraction enterprises alike, the governance of developing societies underwent a sea change during the 1980s (Martinussen 1997). The impetus to involve ordinary people in project design and implementation arose from a rethinking of the politics of development, to favor greater participation for its own good and to reorganize the power relations between expert and beneficiary (Chambers 1997). Yet as this was occurring under the umbrella of neo-liberal restructuring of governance, participation was appropriated into measures to provide technically efficient policing practices through which the relationship between state, civil society and the market could be recast (Cooke and Kothari 2001). As with any aspect of development, participation and forms of governance remain highly contested and politicized (Hickey and Mohan 2004; Watts, this volume).

The concept of social capital – that is, embedded meaningful forms of social organization and networks – has become widely used in the development field, especially by economists and neo-liberal development policy-makers (Fine 2001b). Social capital appears to offer a linking concept, a concept around which conversations can occur between economists, political scientists, and anthropologists (Bebbington 2002; Watts 2002). Social capital models aim for economic development via the encouragement of a robust civil society and social trust. Although referring primarily to social relations, social capital is at times equated with culture as if they were two sides of the same coin. Culture, in the words of a development economist, comprises "all dimensions of social capital in society, underlying its basic components" (Kliksberg 1999: 88). By coming close to a culturalist explanation, these frameworks risk treating social capital as a "black box" by celebrating visible elements of the sociocultural without examining the power relations, silences and exclusions upon which they are erected (Radcliffe 2004). In this context, social capital risks being treated as separate from the political, racial, and contested social relations that contextualize it, just as in the broader concept of culture from which it gains so much power. Where social capital is tied much more closely to the notion of organizational cultures, it may offer a greater insight into the types of social relationships that underpin successful development (Staber 2003). The conversation initiated by social capital would be usefully directed towards the analysis of class, gender, and racial-ethnic relations that underpin social networks and forms of inequality. The challenge is to maintain this conversation around culturally meaningful interactions embedded in complex articulations of politics, power, race, and global-local ties (Porter and Lyon, this volume; Nederveen Pieterse, this volume).

Culture and development in context: an introduction to the chapters

Drawing on the above discussion about development and culture (and its contributing debates), the chapters in this volume examine the dynamics of socioeconomic change and development projects in a number of different contexts around the world. The chapters address the question of *how* specific development projects and broader socioeconomic processes engage the diverse applications of (diverse notions of) culture, while identifying processes that might provide insights into *where* and *when* culture is associated with beneficial development outcomes. Development's cultures are not at a remove from the market and the state, but constructed at the interface between political economies and racial formations, in relation to a combination of global, "Western," national, and local agendas. Understanding culture and development thus necessitates the analysis of the specific *content of development's cultures* that prevail at a particular time and place. Culture has to be situated firmly within an analysis of development's grids of power where hierarchies between world regions, races, cultures, and modernity/tradition act to differentiate social actors. Culture has always been present in development, but has been defined, managed, and fought over in diverse ways at different times and in different spaces. Accordingly, development's cultures are defined and negotiated in a multicultural field, constructed in the unequal interactions between contingently defined groups (Mitchell, D. 2000). Development is intrinsically a field of social interaction between multiple conceptions of culture, tradition and modernity.

The chapters that follow show how different understandings of "culture" have been taken on board in development thinking in specific contexts and development projects. In addition to describing the way that culture is conceptualized in each case, the chapters analyze what difference the new "paradigm" of culture and development – in all its regional and local varieties – makes on the ground. Questioning the assumption that culture is a simple or singular factor to be "added in" to development thinking, the chapters examine the ways in which culture and development thinking plays out in relations of power, inequality and strategic essentialism while having uneven impacts on groups' livelihoods and empowerment. By means of theoretically grounded and substantively detailed accounts, the chapters provide a critical examination of the complexities and unintended consequences of taking culture seriously in development. They highlight the ways in which culture comes together with development across multiple scales and in relation to diverse actors, institutions, and imaginaries of desirability.

Michael Watts' chapter provides an overview of development's history since 1945, and how culture has played a constitutive role in development changing over time with shifting development paradigms and varying expectations about the cultural economies of developing societies. Development theory has always borne the traces of culture and a theory of the modern. Postcolonial theory and the new historicist studies of development theory have been especially attentive to such cultural traces, to what Charles Taylor calls a cultural theory of the modern. After modernization theory's assumption of cultural convergence to culturalist accounts, current approaches to cultural development are explored in terms of how they reframe three key arenas of development: the state, the market, and civil society. Drawing on a cultural theory of the state, the market, and civil society, Michael Watts focuses specifically on the ways that concepts of community and scarcity underlie much of the debate around cultural development today.

Elizabeth E. Watson's chapter departs from examples of projects on the ground in Ethiopia where the cultural "resource" of existing sociocultural relations, authority, and meanings has recently been reinserted into development thinking and practice. Arguing that participation paradigms draw on specific understandings of culture, Watson shows how, in the case of natural resource management projects in Ethiopia, "traditional" institutions have been placed at recent projects to establish management of people, places, and resources. The chapter carefully traces the changes in development thinking that lead to these projects, and then evaluates the ways in which the projects work out on the ground in the context of a post-conflict society living with a new constitution and urgent environmental issues.

Indigenous people in Latin America have long been perceived by their co-nationals and external observers as having distinctive cultures, yet due to their political and sociocultural marginalization from national development processes in the region, these cultures have often been treated as historic relics that would disappear with modernity. Sarah A. Radcliffe and Nina Laurie's chapter shows how, by contrast, recent development thinking in the Andean countries has begun to treat indigenous cultures as an asset rather than a burden, responding to indigenous political mobilization and changing patterns of state rule. However, indigenous empowerment and livelihoods are not guaranteed by recent policy, in part due to the discursive and practical restrictions placed on indigenous places, activities, and relationships with other places and groups.

Paradigms of Andean indigenous development rely heavily on the assumption that Indian groups have high levels of social capital but lack other forms of capital. Two chapters here engage critically with the concept of social capital, which has underpinned several culture and development policy initiatives and development thinking in many areas of the global South. Jan Nederveen Pieterse's chapter asks how social capital relates to cultural difference, questioning the standard account of social capital as being culturally bounded. He furthermore unpacks the notion of "ethnic economy," thereby furthering the anthropological deconstruction of the concepts of culture and race (see above). By unpacking these notions, he comes to focus on the cultural practices and cultural economies by which immigrant enterprises become established, not least by working *cross*-culturally whereby "immigrant economies are embedded within intercultural economies" (p. 141). He ends this chapter with policy considerations, stressing the importance of intercultural communication, and access to capital of various kinds.

Also furthering our understanding of the cultural implications of social relations is Gina Porter and Fergus Lyon's chapter. Drawing on extensive fieldwork with a variety of development initiatives in Ghana, Porter and Lyon explore the role of civil society groups in development and critically engage with the development studies literature on social capital. While the latter emphasizes the virtues of citizen groups (in generating trust and empowerment), local development practitioners highlight the role of culturally defined sanctions and shame in maintaining group cohesion. The chapter goes to the heart of recent debates about the role of civil society institutions in the delivery of development. Arguing that recent World Bank interpretations of social capital equate it with group cooperation, the chapter details two grassroots projects where this interpretation of social capital has guided development practice. The chapter concludes that beyond the identification of social capital as "groups," development policy has not grappled with the historically and geographically nuanced relationships in society that give rise to different degrees of cooperation and development success.

The collection's focus then shifts to places outside the global South, away from the areas predominantly associated with development concerns and policy interventions. Yet as this introduction highlights, the debates around capitalist market development and the experiences of East Asian cultural economies have blurred the boundaries between development and other disciplines. The musicians described by Michelle Bigenho work to bring Bolivian music to Japanese school children, exemplifying the kinds of culturally distinctive services and products mentioned in Sarah Radcliffe's and Nina Laurie's discussion of Andean indigenous development. What Bigenho makes clear, however, is how Japanese expectations of authentic Bolivian culture are produced and reproduced in a multiethnic and multicultural interaction (which includes Bigenho herself playing fiddle). Under late capitalism's culturalization of economic life, the music group meshes well with a Japanese cultural economy that values the Bolivians' flexible acceptance of lengthy touring contracts, multiple skills, and the ability to embody cultural difference. Critiquing celebratory accounts of cultural entrepreneurs, she shows how the labor involved in music performances involves a lot of hard work that permits small gains in living standards "back home" and no simple access to global stardom.

In the context of policy promoting Silicon Valley-like clusters of dynamic firms, the example of the Mormon region in Utah, in the United States, offers a prime location from which to explore in depth the impacts that regional economy have on the dynamism of capitalist economic sectors. In his chapter, Al James unpacks the notion of cultural economy by exploring how and in which practices regional culture conditions firms' innovative behavior and economic performance. By exploring the interactions and overlaps between regional, firms', and industries' cultures, James demonstrates how Mormon regional culture both enables and constrains innovation.

The last substantive chapter returns to the majority world, specifically to South Africa, where cultural development must be seen in the context of post-apartheid settlements and tensions. In South African majority rule, various efforts to revalue specific cultural practices have been made. For example, legislation to recognize the skills of around 200,000 traditional healers or *sangoma* has been introduced (Díaz 2005). Cheryl McEwan's chapter addresses how the memories of apartheid violence are being reworked culturally by women's embroidery, in order to claim a new voice in development decision-making and practices. In situations where civil war and extended periods of violence against subsections of populations have destroyed social patterns and disrupted cultural meanings, culture has been pinpointed as a potential bridging activity across tense social divides. Having the necessary security and peace to undertake a cycle of ritual events indicates the end of hostilities in a civil war, yet in many cases the meanings around rituals will have been transformed by experiences of violence and disruption.

In the context of rising global inequalities and the pervasiveness of culturalist international politics that pit "the West" against "Islam," culture represents a urgent arena through which development issues and the geopolitics of development assistance can be considered. While culture is far from being a panacea, the insights to be gained by taking it seriously in development thinking and practice offer new horizons for the world's culturally and materially disempowered people.

Notes

1 Mozambique and Peru were early adopters of UNESCO's recommendations, both being countries with recent histories of destructive civil strife and where particular policy attention was awarded to social cohesion and poverty reduction (Davis 1999). In Mozambique, bounded (ethnic) cultures were the basis of post-conflict cohesion building on what was assumed to be a pre-existing organizational culture (see Black and Watson 2006). In Peru, development had to address indigenous peoples' historic marginalization in racist rural development (compare Radcliffe and Laurie, this volume).

2 The persistence of culturalist explanations in everyday life rests upon precisely this quality. Hence, in the case studies that follow, authors may identify culturalist explanations among development actors on the ground yet their own understanding of the concept is quite different.

3 This perspective has not disappeared; see Kuran, who argues "cultural barriers to material growth, where they exist, are generally neither unalterable nor insurmountable" (2004: 120).

4   Compared with accounts of development that downplay the hierarchies of
     multiracial society, Third World feminism addresses race-ethnicity and gender
     as key dimensions of social difference, shaping development outcomes through
     their interaction (Mohanty, Russo and Torres 1991; Alexander and Mohanty
     1997; Chua, Bhavnani, and Foran 2000). Challenging the racial and cultural
     hierarchies that marginalize women and ethnic-racial populations involves a
     critique of the power relations in international development and how they play
     out at national and local levels, as well as a recognition of how these relations
     are constructed in multiethnic and multiracial settings (Bhavnani, Foran, and
     Kurian 2003).

5   By treating culture and economy as separate, the concept of cultural
     embeddedness implies that immanent economic laws are played out in the
     cultural realm, or that culture is mobilized for economic gain (Amin and
     Thrift 2004: xiv).

2 Culture, development, and global neo-liberalism

Michael Watts

[L]arge related bodies of thought appear, at first like distant riders
stirring up modest dust clouds, who, when they arrive, reproach one
for his slowness in recognizing their numbers, strength and vitality.

(Glacken 1967: xi)

In every culture there is something that works, and the thing is to find
out is what that is. Is it the headman, the religious leader, community
pressure, or the police? Find out what it is and use it.

(USAID operative, Cairo, 1995, quoted in
Elyachar 2002: 509)

How did it come to pass? The notion – captured perfectly by the
USAID trainer offering micro-entrepreneurial counsel to Egyptian
NGOs – that *culture* is the indispensable stuff of development. The
museum of culture is to be ransacked by the development
practitioner (and by implication the development theorist) in search
of things that "work." In the name of development, culture must
be instrumentalized – what we might more properly call the
"economization of culture." At stake is yet one more expression
of the colonization of the life world by the deadly solicitations of
the market. In Elyachar's example, the headman becomes the
enforcement mechanism for Grameen banks Egyptian-style. The
pursuit of some cultural form capable of being put to work presumes,
however, another aspect of the complicated dialectics of economy
and culture. Namely, the idea that economic practices are necessarily
culturally "embedded"; the economic is represented (or materialized)

through the symbolic (what we might gloss as the "culturalization of the economy").[1] Here economic determination is reversed; markets cannot function without culture. The dull discipline of the market requires its own cultural conditions of existence: to operate, in other words, the market requires trust, networks, norms and values, and various institutional prerequisites including law.

How is it, then, that the Berlin Wall separating economy and culture has collapsed – or at the very least, the boundaries between them are ever more blurred and porous? Lash and Urry put it this way: "the economic and the symbolic are ever-more interlaced and interarticulated . . . the economy is increasingly culturally inflected and culture is more and more economically inflected" (1994: 64; compare James, this volume). These inflections are apparently as compelling for the development establishment – the US Agency for International Development and, as we shall see, the World Bank – as they are for the purveyors of contemporary social theory. And who were culture's "distant riders" (to invoke Clarence Glacken's magnificent turn of phrase) who battered down the walls and punctured the boundaries, the horsemen whose numbers, strength, and vitality now surprise us? These are the questions I should like to briefly reflect upon in this chapter.

To recognize the centrality of culture in the world of development at this moment – in the realm of ideas and institutions, that is to say as system and practice – is, of course, to pose a paradox of sorts. Let us recall that the Reagan–Thatcher–Kohl neo-liberal agenda was already in high gear by 1985 and the IMF–Wall Street–Treasury complex had succeeded, to a remarkable degree, in not only discrediting the old Keynesian doctrines but in delegitimizing anything that might inhibit the free movement of capital. "Getting the prices right," global competitiveness, the new realism, shock-therapy, freedom through trade, "there is no alternative": these were the heady prescriptions screamed by the multilateral circus barkers from the World Bank and the International Monetary Fund. What they celebrated was not simply the prospect of a "neo-liberal grand slam" (the language is taken from Perry Anderson [2000]), but more profoundly a new kind of universalism – an epistemological counter-revolution – at the very moment that postmodernism claimed master-narratives to be dead. The *laws* of economics, said then Clinton Treasury Secretary Lawrence Summers, are "like the laws of engineering"; there is only one set and "they work everywhere." So why has the rising tide of neo-liberal orthodoxy brought with it a flotilla of cultural imperatives? Even within the heart of the World Bank there are now voices for whom development is "getting the social relations right" (Woolcock cited in Harriss 2002: 81). The paradox is that the cold, universal calculus of free-markets – the (universal) instrumental logic of mean and ends (Sayer 1997: 17) – at its moment of triumph has spawned a cultural counter-revolution: a widely held recognition that the market economy is no less a site of culture than the household or the mosque. The "monologic" of the economy can never escape the "dialogic" of culture.

None of this is to suggest that the hard-nosed economists in the structural adjustment division of the World Bank are now fully immersed in Jacques Derrida (but there are certainly program and operations officers who are sufficiently self-reflexive to incorporate the latest ethnographies of development into their arsenal [see Li 2005]). Culturalist thinking, nevertheless, has scaled the walls of the most orthodox of development institutions.[2] And it has done so against a backdrop of momentous – and intuitively rather inhospitable – change in the policy climate (Duggan 2003). To have proposed in 1945 the sort of conservative revolution we have witnessed over the last two decades – predicated as it is upon appeals to science and reason to endorse the radical application of the laws of the market – would have been the quickest way to the mad house (Giroux 2004). Yet this passive revolution from above – made in the name of a ruthless American hegemony – has rekindled the fire of culture. One only need think of three powerful bastions of development convention: the transnational corporation (TNC), the United Nations Development Program (UNDP), and the World Bank.

TNCs and corporate social responsibility (CSR)

Corporate culture – the "ensemble of material practices, social relations, and ways of thinking" (Schoenberger 2002: 378) within a corporation – is as much a part of the business world as it is an academic object of scrutiny. But there is another cultural arena in which the conduct of TNCs has generated an entirely new policy arena: corporate social responsibility (or corporate citizenship) (Hopkins 1999; Smith 2003; Kotler and Lee 2004). The World Business Council for Sustainable Development defines CSR as: "the continuing commitment by business to behave ethically and contribute to economic development while improving the quality of life of its workers . . . as well as the local community and society at large" (World Business Council for Sustainable Development, 2005). At one level CSR is nothing more than good business sense, endorsing corporate policies that any company should have in place (not lying to employees, not paying bribes, conforming with national labor legislation, and so on). But in its most ambitious form CSR attempts to make binding commitments upon companies to ethical investment and to embed corporate activities – as a locus of enormous non-state power – more fully in the architecture of the UN Universal Declaration of Human Rights, the UN International Covenants on Civil and Political Rights and on Economic, Social and Cultural Rights, and the UN Codes on Conduct for Law Enforcement and basic principles on the Use of Force and Firearms. Modern CSR was born during the 1992 Earth Summit in Rio as an explicit endorsement of voluntary approaches rather than mandatory regulation. What moved CSR forward in the 1990s was a combination of corporate disasters (Shell and the Ogoni, for example [Okonta and Douglas 2001]) and the growing role of the human rights advocacy network. In North America it was the anti-sweatshop movement, the anti-dams movement, and efforts to make the extractive sector more transparent and accountable (the case of "blood diamonds" is a key case) that proved to be the catalysts for the proliferation of corporate codes of conduct. Virtually all major TNCs, for example, now have some code of conduct, and have revamped (on paper at least) their approaches to community development, environmental responsibility, labor relations, and corporate conduct (Watts 2005). Much of this remains voluntary and what one might call "para-legal" (the codes often have limited enforcement). Nevertheless, the rise of CSR has moved hand in hand with new corporate development initiatives operating under its banner. Community development – "sustainable community development" is the current moniker – places corporations in the business of doing development, working with NGOs and development agencies, and hiring rafts of anthropologists and development practitioners. It is not so much that TNCs (or development institutions for that matter) have their own culture but rather a parallel privatization of development through community initiatives that has thrust TNCs into the complex territory between economy and culture (Prahalad and Porter 2003; Sullivan 2003). Corporate practice depends upon – and aspires to build (or deepen) – those sociocultural relations that constitute community. One can, of course, be critical or suspicious of its CSR's efficacy but nobody should mistake the extent to which corporations are in the business of making and breaking culture as a way of making development good for shareholders.

UNDP, democracy, and human rights

One of the striking new developments within the world of multilateral developments institutions is the extent to which we have witnessed, over the better part of 20 years, two rather different structures of knowledge for discussing, measuring, and assessing development. One is the World Bank's annual compendium (the *World Development Report*) of statistical data typically organized around a policy theme – the environment, service provision, inequality. The production of these reports is complex and multifaceted, and often contested as the infamous (and public) debate over the 2000 *Attacking Poverty* report revealed. But in general it is clear that the Bank privileges national accounts data and a raft of conventional economic measures (GNP, GDP, and so on). The measures (World Bank Indicators) are typically deployed in normative terms to serve the interests of a particular vision of development in which free markets and economic growth figure centrally. The forms of developmental classification are conventionally around income (low, middle, high, and so on). None of this is to suggest that the measures and concepts are static – the introduction of purchasing power parity is a case in point – or that getting the prices right always crowds out other concerns like governance or environment, but the commitment to the neo-liberal model is surely incontestable. The UNDP *Human Development Report* conversely has emerged as a sort of ideological and conceptual counterweight to the Bank's annual inventory. Its central index is not GDP per capita but the Human Development Index and the Human Poverty Index – both of which depend upon rather different measures beyond income or market-valued output. Here it is longevity, knowledge, standards of living, and social exclusion that matter. The rankings that are of consequence are the anomalies between expected development achievements based on GDP and actual human development.

If the theoreticians and stenographers of the Bank are the partisans of the neo-liberal counter-revolution, the UNDP, one might say, has been captured by Amartya Sen (1999), the women's movement, and the NGO world. UNDP highlights entitlements, capabilities, and freedoms as its preferred building blocks; its concerns are deepening global inequality, the relations between human rights and development, the failures of markets, the need to build safety nets (consolidating and deepening entitlements) through empowerment and power mobilization. Nowhere in the Bank compendium is one likely to see – as is the case in the *Human Development Report* – tables depicting who owns the media, or the relations between democracy and equity, whose voice counts at the World Bank and the IMF, corporate influence on policy, and the relations between gender and legal process. The UNDP's ideological vision is in no simple sense Left; it is perhaps Polanyian. The notion that left to its own devices unregulated markets are massively corrosive and their tendency to "disembed" (to commodify all aspects of social life) will always throw up institutions and movements to protect society from its ravages (Polanyi 1944, 1945; Burawoy 2003). My point is that UNDP represents one enormously influential trend whose vision of development – and means for measuring it achievements and failings – are culturally inflected.[3] This is true in at least two senses. First, is that development is always culturally constituted (UNDP's admirable commitment to minority rights, gender justice are cases in point). And second, what the twin UNDP–World Bank programs represent are not just different ideas but a struggle between the ways in which the world of development practitioners is shaped by particular systems of knowledge. One can do no better than quote cultural anthropologist James Ferguson from his ethnography of an African development project: "what [development bureaucrats] do and do not do is a product not only of the intense interests of various nations, classes and international agencies, but also of the working out of . . . [a] complex structure of knowledge" (1994: 18). Development as a structure of knowledge is, broadly construed, a cultural question – or perhaps one should say a cultural achievement (or a form of hegemony).

The World Bank and social capital

What began as a meditation on Italian democracy and on the decline of the American bowling club (Putnam, 1993; 2000) now has its own glossy location on the World Bank website. The meteoric rise after 1995 – the "big bang" as Harriss calls it (2002: 75) – of social capital as the "missing link" of development is one of the most striking aspects of contemporary development theory and practice. Two books (Fine 2001b; Harriss 2002) have already appeared documenting its institutionalization within the World Bank, and its ability to travel far and wide within conventional development practice. Social capital should be seen as the twin sister of another development discourse (also recent in provenance and now an important part of what is called the "post-Washington Consensus"), namely governance. Both are products of the rise of civil society, especially in the wake of 1989, as a development arena and of the growing recognition that corrupt and unaccountable states could compromise the purported benefits of "emerging markets." The story of social capital is how a notion with a long history within the academy during the twentieth century (and exhibiting more or less radical iterations), and fed by strong communitarian currents in contemporary philosophy, found a receptive audience within the World Bank. which in turn lent its authority and legitimacy to an idea – "getting the social relations right" – that is now nothing less than a small-scale research industry. For some entrepreneurs within the Bank, social capital was a Trojan horse; it represented an opening, a way of sneaking in a challenge to a fundamentalist commitment to markets as an end in themselves (without a recognition that markets do not automatically give rise to the "right" institutions and getting markets to work better requires something more than price setting). Social capital, on this account, attempts to "socialize" the economists: "The broad and most encompassing view of social capital includes the social and political environment that shapes social structure and enables norms to develop" (http//www. worldbank.org).

In the Bank's lexicon, social capital can "bond," "bridge," and "link." Norms, networks, trust, forms of association . . . this is the raw material of social capital which, it turns out, is key is keeping government honest, to making institutions work, in securing access to markets. Social capital confirms "a growing body of evidence that incorporating the poor into the design and implementation of . . . projects" (World Bank cited in Harriss 2002: 94). In turn, social capital – networks, associations, norms, and values – must be identified, used, invested in, and enabled. There are perfectly good reasons for assuming that the critical tradition within the genealogy of social capital – the work of Pierre Bourdieu (1980) or the research of Peter Evans (1996) to take two examples – has been evacuated or lost as it is has been "domesticated" within the Bank. What is lost sight of is the wider process of democratization within systems of structured inequality and power that determine whether social capital and community development is anything more than another palliative, another "anti-politics machine" (compare Porter and Lyon, this volume). But social capital turns, nevertheless, on a cultural breaching of the wall that surrounds the economy. Culture again raises its head. Norms, values, associations: we may differ as regards their capacities in the way the Bank envisages their deployment. But their status as culture is unassailable.

*

These three brief illustrations are drawn from what we might call the bastions of orthodox development practice. If we turn instead to the enormous ocean of non-profits, social movements, and civic groups operating under the banner of Port Alegre – including the so-called "movement of movements" (Mertes 2003), and the motley transnational community of activists, academics, and public intellectuals that collectively constitute the "post-development" or "alternatives to development" school (Rahnema and Bawtree 1997; Nederveen Pieterse 2001b) – then the shift away from a crude economism is clearly complete. Two geographers – Dennis Cosgrove and Peter Jackson – penned something like a manifesto for cultural geography in the mid-1980s in which culture was no longer residual, it was the "very medium through which change is experienced, contested and constituted" (1987: 99). From the vantage point of the new millennium, there is nothing here that would surprise any development specialist. Culture, one might say, has become "normalized" within large swaths of development discourse. So how did culture emerge in these ways, how and why did it envelop development and with what consequences?

One must immediately note, of course, that there is an important sense in which development has always – and sometimes self-consciously – been construed in cultural terms. The very notion of ending poverty was a product of the late eighteenth century (Stedman Jones 2004) – the confluence of scientific progress, the Revolution in France, and the promise of the new international economy – and it was debated (and ultimately defeated) in cultural terms. The moment at which such a notion was extended to the non-West in the 1890s as part of a colonial project – what Cowen and Shenton (1997) call "trusteeship" – was wrapped up with the disenchantment associated with capitalist destruction (it emerged, in sum, from the contradictions of modernity). What was on offer then – and again in Truman's 1949 invocation of "a program of development . . . based on democratic fair dealing," or the postcolonial energies unleashed at the Bandung Conference – was a sort of utopian vision based, as Alexander Gerschenkron once noted, on the need for a "New Deal of the emotions" (cited in Watts 2000). If culture was present, so to say, at development's inception, it has retained its historical appeal even in the face of the high economism of development theory. Colonial development, after all, was in large measure a cultural project. The British and French in their African colonies governed through culture ("decentralized despotism," as Mamdani 1996 calls it), they fretted over the "tribal" problem and the social disintegrative effects of urbanization, they rooted improvement in custom and tradition, and sometimes even came to value local peasant knowledge and practice. Even within the "age of modern development" in the wake of World War II, culture was not overrun by the economists and engineers armed with slide rules and talk of savings rates and capital accumulation. Modernization theory in the 1950s was self-consciously cultural. Not simply because its founding text – Rostow's non-communist manifesto – was an unashamed projection of the West as a global norm, but because its conceptual tool kit latched on to culture as measure of backwardness: how else to explain the "culture of poverty," the mountains of paper devoted to explicating peasant irrationalities, and the phalanxes of consultants put to the task of inculcating "need achievement" among the natives (see Cooper 2005)? And not least, the first generation of postcolonial nationalists sought to *indigenize* development for their own purposes, tailor-made for local conditions: Nyerere's *ujamaa* rooted in African traditions, Nehruvian socialism, socialism with "Chinese characteristics," to say nothing of the cultism of Mbutu or Sun Yat Sen.

What I am calling the "cultural revolution" in development – and what has elsewhere been called the "cultural turn" in the social sciences (Jameson 1998) – marks something more than a sort of latent recognition of symbolism, meaning, and ideas (which is to say culture showed up in development theory as modernity's other: tradition). One way to grasp this shift is to start with the recognition that development theory and practice turns on discursive – by which I mean systems of meaning and practice – and normative battles over the weighting of states, markets, and civil society in the design of human improvement. The new President of the World Bank (Paul Wolfowitz), Amartya Sen, and the leaders of MST in Brazil represent quite different positions with regard to, say, the centrality of free markets or state regulation in development. If each leg of this tripod – crudely put polity, economy, and society – constitutes the stuff of development theory and practice, then the power of the cultural revolution is reflected in the extent to which the meanings of each of these three spheres has been "culturalized."

Culturalization is something more than Sewell's (1999) observation that in the last three decades or so culture has emerged as an autonomous realm; it is that the economy (to take one example) has been construed in cultural terms (the notion, with which I began, that the economy-culture boundary has been punctured). But the same holds for the state – the rise of cultural politics, of the politics of recognition is a case in point – and associational life. One measure of this culturalization is the fact that each realm has been explored ethnographically (see Mitchell, T. 2000; Hansen and Stepputat 2001) – the methodological hallmark of cultural study; another would be the extent to which (and I shall return to this point) development is seen now as a form of modernity (and often as a catastrophically *failed* modernity). There is no simple way in which polity, economy, and society are rendered in cultural terms. Crang (1997) properly notes that the economy–culture binary has been (re)read as: an economic determination of culture, an economic operation to culture, an economic colonization of culture, an embedding of the economic in the cultural, a representation of the economic through the cultural, or a cultural materialization of the economic. What matters, however, is that the cultural turn has indisputably colored – colonized might be a better term – all the human sciences.

Culture in development

To posit a cultural revolution begs the question of what exactly it is that has colonized development discourse and practice. Definitions are in order but one runs up immediately against a word of formidable complexity. It has been said, with good reason, that nature and culture are perhaps the two most complex words in the English language. They carry, as Raymond Williams (1976), once said, an enormous amount of human history. Taken together the two words are often assumed to be opposites – the material and the ideal, the biological and the semiotic, a realm of law and a world of contingency – but on closer examination their polarities are tangled, difficult, and intractable. As Terry Eagleton puts is, culture inherits the imposing mantle of religious authority but also has "uneasy affinities with occupation and invasion" (Eagleton 2000: 2). Culture is obviously not reducible to nature, but their referents are not stable either. They are shifty, tangled, and mixed up; antinomies seem to proliferate. Culture may be nature's other but there is much traffic, a veritable information highway, between them. Culture is built out of the "ceaseless traffic with nature which we call labor" (Eagleton 2000: 4; see also Bigenho, this volume).

How might we grasp culture's historical semantics and its relation to contemporary development? Raymond Williams (1976) has charted three major modern senses of the term culture: as *utopian critique*, as a *way of life*, and as *artistic creation*. The first must be situated against the horrors of industrial capitalism, as a sort of anti-capitalist impulse. The word first comes to mean civility and later civilization (understood as progress), but around the turn of the nineteenth century the descriptive and normative aspects of the word, as Eagleton says, "fly apart" (2000: 10). Civilization appears increasingly debased, predatory, and culture appears as a response to "radical and painful change" (Williams 1993: 60). The second shift marks culture as a way of life and is implicitly a critique on the universalism of the Enlightenment. Culture is rendered multiple to encompass, for example, the Romantic anti-colonial desire to reclaim the exotic and the traditional in the face of industrial triumphalism. Cultures are organic and rooted and could furnish, in the hands of Johann Herder or Edmund Burke, a radical critique of Eurocentrism or the notion of industrial capitalism as universal civilization. And third, Williams charts culture as a gradual specialization to the arts whether broadly construed as intellectual pursuits or science, or more narrowly confined to literature and the imaginative arts. Culture figures here, above all else, as refinement, impartiality, and "agreeable manners and an open mind" (Eagleton 2000: 180).

In Williams' language, culture is a "binding" word which sutures these differing activities and their interpretation, and is also what he calls an "indicative" word appearing in specific forms or lineages of thought (for example, nationalism or racism). These complex and shifting modern meanings cannot be identified, so to speak, in advance; rather they are philosophical and historical texts, always deployed in particular ways, and inextricably bound up with the problems they are being used to discuss.

Here lies culture's complexity, its perplexing doubleness; it is "both synonymous with the mainstream of western civilization and antithetical to it" (Young 1995: 53), and "both material reality and lived experience" (Eagleton 2000: 36). Culture's polysemic qualities are perfectly captured in a new geography handbook (Anderson, Domosh, Pile, and Thrift 2002): culture, the authors point out, can be a distribution of things, a way of life, a universe of meaning, a way of doing, and a field of power. The multiple meanings of culture are part of its appeal and durability, and its ability to travel into development. Culture can and does mean differing things to differing constituencies.

So how might we trace the lineages of culture in relation to development, at least since the Cold War, to identify meanings that have particular traction in the age of development? This is an enormously difficult question – compounded by the fact that it is discursive, operating at the level of both ideas and texts (theories, policy statements) and practices (the institutional use of cultural ideas). All that I am able to do here is to sketch some broad ideas. A fundamental starting point has to be the sociology of knowledge, and more precisely a wide ranging "cultural turn" in the human sciences over the last 30 years or so (Jameson 1998; Bonnell and Hunt 1999). There is no simple account of this turn (a phenomenon largely of the academies of the trans-Atlantic). Cultural studies in Britain – associated with Marxist critics (Raymond Williams, Stuart Hall) and social historians (Edward Thompson) in the 1950s – struggled with the limits of Marxist economism and the ways in which the concept of class could be retained through notions of "experience," "structures of feelings," and "cultures of class" (Nelson and Grossberg 1988). In the United States, the birth of the turn is typically traced to Hayden White's *Metahistory* and Clifford Geertz's *The Interpretation of Cultures* both published in 1973; in France to Pierre Bourdieu (*Outline of a Theory of Practice*) and Michel Foucault (*Discipline and Punish*) and their canonical work on cultural capital and governmentality published in the late 1970s. The problems and questions to which culture was put in these differing national settings – and the tool kit developed in response to them – was obviously not of a piece: cultural capital here, thick description there; discourse and habitus in one locale, semiotic texts and poetic acts in another. What one can say, I think, is that what was at stake was first, a questioning of the social (as a category and as a style of explanation), and second, a concern with language and interpretation as the means by which social categories came into being.

William Sewell has captured perfectly the broad outlines of what the cultural turn produced:

> [C]ulture should be understood as a dialectic is of system and practice, as a dimension of social life autonomous from other such dimensions both in its logic and in its spatial configuration, and as a system of symbols possessing a real but thin coherence that is continually put at risk through practice and therefore subject to transformation.
>
> (1999: 52)

The effect was to question the established anthropological sense of culture as consistent, integrated, consensual, coherent, resistant to change, and bounded. Rather as worlds of meanings, cultures were unstable, contradictory, loosely integrated, porous, and contested.

I think that there are two especially significant ways in which this constellation of ideas about culture provided a framework for "culturalizing" politics, economy and society, taken from Crang (1997) and Sewell (1999). The first is culture understood as the *meaningful mapping of the world* (worlds of meaning or social imaginaries) *and the positionings of culture bearers within it*. It speaks to identity, meaning, and signification as forms of practice. In the second, culture turns more on difference than identity, that is to say, *the ways in which mappings can produce distinctive social groups and the institutions that are central to their constitution and reproduction* (recognizing that the old sense of cultures as bounded, coherent social units must be abandoned). Inevitably these two broad articulations overlap and intersect in important ways. Both contributed to the idea that the world might no longer be neatly cut up into bounded societies with their own cultures yet retaining a powerful sense in which, as Sewell noted, there are particular consistencies in the ways in which, in particular places and times, worlds of meaning "hang together." I want to turn to each of these broad meanings – which I shall gloss as social imaginary and community – insofar as they help frame the cultural turn in development.

Culture as [developmental] community

> [Community] is not primarily a geographic space, or social space, a sociological space or a space of services, although it may attach itself to any or all such spatialization. It is a moral field binding persons into durable relations. It is a space of *emotional relationships* through which *individual identities* are constructed through bonds to *micro-cultures* of values and meanings . . . in the institution of the community a sector is brought into existence whose vectors and forces can be mobilized, enrolled, deployed in novel programs and techniques

and harness active practices of self-management and identity
construction, of personal ethics and collective allegiances.

(Rose 1999: 172–6)

Deployed in the English language for at least 500 years,
community has carried a range of senses denoting actual groups
(for example, commoners) and connoting specific qualities of social
relationship (as in *communitas*). The complexity of community
therefore relates to the difficult interaction between two historical
tendencies:

> [On] the one hand the sense of direct common concerns; on the other
> the materialization of various forms of common organization, which
> may or may not adequately express this . . . Community can be the
> warmly persuasive word to describe an existing set of relationships,
> or the warmly persuasive word to describe an alternative set of
> relationships. What is most important, perhaps, is that unlike all other
> terms of social organization . . . it never seems to be used unfavorably,
> and never to be given any positive opposing or distinguishing term.
>
> (Williams 1973: 76)

Community spoke to membership and identity in which interests,
property, and shared meanings were at issue. There were signs,
however, from the seventeenth century of a sort of rupture in its
usage – which was to become especially important with the advent
of capitalist industrialization – in which community was felt to be
more *immediate* than society. By the nineteenth century, of course,
community was invoked as a way of theorizing modernity itself (and
here is, as we shall see, a key connection to the idea of culture as a
social imaginary). Community – and its sister concepts of tradition
and custom – stood in sharp contrast to the more abstract,
instrumental, individuated, and formal properties of state or society
in the modern sense. A comparable shift in usage, which has
occurred in the twentieth century, was also noted by Williams in
which community came to be invoked as a way of discussing a
particular sort or style of politics distinct from the formal repertoires
of national and local politics. Here the reference is direct action,
direct community participation, and organization embracing,
typically, a populist notion of working with and for "the people."

One can recognize immediately how culture as community has a
deep presence in development practice (whether capitalist or
socialist): Indian cooperatives, native administration, the commune

(Kitching 1980). In its contemporary iterations, however, it is the empowered and self-governing community that has the deepest resonance. Empowerment represents one major way in which the community has entered, or perhaps re-entered, development discourse but along two rather different trajectories. One emerges from anti-systemic movements. Much of what now passes for grassroots anti-development initiatives, or the new sorts of anti-system or ant-globalization movements, starts from the notion of a radical empowerment from below (Mertes 2002). Another iteration of community as empowerment appears within more conventional development circles: for instance corporate community development, social capital, and community-based resource management (see Watson, this volume). Its source is neo-liberal theory itself. Pre-existing communities must be enabled and enhanced in their institutional capacities in order that they can assume the responsibility for their own *self-improvement* by tapping market power, conducting themselves in a competitive arena, and acting in a calculated manner. Both the community, and the civil society of which it is part, must be subject to the rule of the development expert and rendered through technical knowledge (Mitchell 2002). The vast explosion of NGOs and the civil society boom of the post-1989 period, represents a vast new cultural terrain on which donors and governments now operate in the name of "enabling environments" (Li 2005).

There is another important line of communitarian theory within development that centers on the now vast body of work on nationalism. The idea of "new nations" and development was, of course, central to 1950s modernization theory but it was Benedict Anderson's *Imagined Communities* (1983) that marked a brilliant rethinking of the nation around the social imaginary and cultures of community (another point of connection between the two meanings of culture). Nationalism – the imaginary word of horizontal comradeship organized around space – was unleashed by print capitalism (and later the virtual world, what he dubbed fax nationalism). The nation was the cultural community par excellence, a particular "style of imagining" out of which postcolonial development was to be fashioned (Coronil 1999; Goswami 2004; Apter 2005). In the postcolonial world the ways in which certain forms of entho-nationalism won out (often violently) and how such political processes were central to nation-building as an institutional project has had the effect of refiguring what had been until the 1970s a rather anodyne (and Panglossian) view of the creation of the modern nation-state (compare Bigenho, this volume). What was central to this cultural turn turned out to be the shallowness of nationalism in many postcolonial settings, and how particular development regimes failed in their imagining of what Laurent Berlant (1991) called a "national fantasy," and conjured up in its place a crisis of secular nationalism – that is, a profound sense of the bankruptcy and failure of modernity. In some cases the very idea of state building and national identity imploded – Somalia, Congo Brazzaville, Afghanistan – while in others, different styles of development – Islamism is the obvious case – substituted for the catastrophic failure of the modern nationalist development project. What is distinctive, of course, of the current moment is that an imperial power – the US – has embarked upon a process of forced nation building imposed militarily from without. Messrs Feith, Wolfowitz, and Rumsfeld discovered that their full-spectrum dominance and the export of democracy – Operation Enduring Freedom and Operation Adam Smith – ran straight into a ferocious wall of cultural politics (RETORT 2005).

Community, in other words, pays fidelity to its modern political usage and at the same time must be located with respect to modern capitalisms. Community is an expression of modern rule (it is in the business of disciplining the liberal subject) but it is simultaneously, as Joseph (2001) puts it, a "supplement to capital" (it is *in* business). There is no simple genealogy to the modern deployment of community, either in the advanced capitalist states, within development discourse, or within the global South. It has come to provide a common currency across the differing expressions of modern justice as Nancy Fraser sees them: the egalitarian *redistributive* sense of justice on the one side (communities of class or social strata) and the politics of *recognition* or recognitive justice on the other (communities of identity and difference) (see Fraser and Honneth 2003). But throughout the nineteenth and twentieth centuries the idea of the community was "a fundamental political institution within European colonial systems" too (Kelly and Kaplan, 2001: 5). Community is an integral part of the construction of the modern. Communities, not just of the modern imagined nation (Anderson 1983) but all manner of local communities, are "political" and "represented" communities to be read against the modern state, the nation, and history. Communities demand visibility, legibility, and enumeration as a precondition for claims-making and thereby entry into the modern – and the modern world of development (compare Radcliffe and Laurie, this volume).

It is, of course, a defining feature of modern capitalism that its competitive and ceaseless search for profitability unleashes periodic waves of "creative destruction," and round upon round of uneven development, through which communities are both destroyed and remade (Hart 2003b). Marshall Berman's magisterial account of the relations between modernization and modernity starts precisely from the *contradictory* experiences of being modern, its vitality promises "adventure, power, joy growth, and transformation of ourselves and the world" and yet "threaten to destroy everything we have, everything we know, everything we are" (1980: 23). Over the last two centuries much utopian populist thinking – often but not always draped in nostalgia for a community lost – can only be grasped as a counterweight to the destructive consequences of industrial capitalism and what passes as development. Working-class communities in Delhi, peasant systems of common property management in the Andes, the moral economy of Islamic schools in Nigeria, and so on, are crushed by the unfettered powers of the market and yet these selfsame conditions of destruction provide a fertile soil in which the endless search for alternatives can take root and flourish:

> The construction of collective identities arises out of broader practices of defining and delimiting communities. As a rule . . . these dissolve internal differentiations within any given collectivity in favor of a common external demarcation . . . The spread of such claims is only possible in communities where religious (or traditional) norms and affiliations have become shaky or uncertain. Surrogate constructions then offer magical formulae that suggest hidden ways of belonging, delimiting and persisting . . . Formulaic constructions of collective identity have become a symptomatic signature of the present. They are ubiquitous wherever societies, regardless of their actual differentiation, are transfigured into seamless communities, and assured of continuity by symbolic demarcation and fabrication of meaning.
>
> (Niethammer 2003: 80–3)

In the making and remaking of *communitas*, whatever its local or historical circumstances, the new community must always address questions of representation (how they represent themselves and what forms of political representation they hold to), forms of rule, means of internal discipline, membership and "purity," styles of imagination and their relation to accumulation and the economy.

Within the maelstrom of capitalist modernity – and global development – the possibilities for community making are almost endless. At this historical moment we are awash in communities, and the "self-governing community" is one of the defining articulations of neo-liberal rule (Osborne and Rose 1998; Rose 1999; Schofield 2002). Li's brilliant new work (2005) properly notes that the community resurgence of the 1980s and 1990s in the trans-Atlantic economies is now far wider – from community policing in Birmingham to community partnerships in the Niger Delta – and cannot be grasped outside of the collapse of socialism and the rise of what Keane (2003) calls "turbocapitalism." Here a new generation of communities arises from the ashes of state withdrawal and speaks the name of civic renewal, associational democracy, and empowerment. Communities are, therefore, bound up with modernity but they are also complicit with modern capitalism itself, and rooted in the operations of the marketplace. It is in this sense that the entry of community into current development discourse must be evaluated.

And yet paradoxically community is an exemplar of what Ernesto Laclau (1996) calls an "empty signifier" – something that no political actor can claim to hold the truth about for very long. For these reasons, three aspects of the developmental community should be clear: first, the fact that community making can fail (often dramatically) by which I mean that a particular world of meaning disintegrates (fails to maintain its social appeal, its ideological function, and its social cohesiveness), and erodes to the point where a base coherence dissolves. The famous Ogoni movement in Nigeria is a case in point. Community can fail to be defined and stabilized (Jensen 2004). Second, that communities typically contain both reactionary (despotic or disciplinary) and emancipatory (liberatory) expressions that are, as it were, in perpetual struggle with one another; communities are not always warm and fuzzy. Development in the name of political Islam is an exemplary case. And third, communities (with their attendant forms of identity, rule, and terrritorialization) can be produced simultaneously at rather different spatial levels and within different social force-fields, and to this extent may work with, and against one another, in complex ways. People, in short, belong to multiple communities (a women's credit association, an ethnic movement for resource control, and a transnational green movement), navigating among them in a way that works against the communitarian presumption that individuals hold fidelity to only one community.

Community is – or shall we say has become – a powerful (but also dangerous) way in which the cultural turn enters development practice and thinking. There are political and material circumstances in which a round of aggressive and destructive global neo-liberalism – let's call it primitive accumulation (Harvey 2002; RETORT 2005) – has been enormously generative for community-based claims making and for what Edward Said called the endless search for alternatives. I shall return to this later. What I have tried to sketch here is the need to retain a critical cultural sense of community in its deployment in development theory and practice: communities are not always warmly persuasive, they can emerge as forms of modern political discipline, they may be cut across by all manner of strong and violent political currents. They can also be the source of generative politics, struggling against and undermining what Ludden (1992: 252) has named "the development regime as an institutionalized configuration of power."

Development as a social imaginary

> Development in other words is Orientalism transformed into a science for action in the contemporary world . . . Postcolonial settings provide the rationale for the idea of alternative modernities . . . where incommensurable conceptions and ways of life implode into one another . . . into strangely contradictory yet eminently "sensible" hybridity.
>
> (Gupta 1998: 45, 219)

Culture as worlds of meaning: what sort of opportunity did this second lineage of cultural theory linking system and practice offer for development? Its most powerful – and most inclusive – iteration is contained within Gupta's twinned observation: that "Development" is a form of Orientalism – that is to say it rehearses "in a virtually unchanged form the chief premises of the self-representation of modernity" (Gupta 1998: 36) – and that postcolonial states rarely swallow the West's self-representation but fashion a hybrid modernity of their own. One of the standard claims of postcolonial theory – a cultural theory par excellence – is that the hybridized postcolonial subjects "continuously interrupt the redemptive narratives of the West" (Gupta 1998: 232). This is a line of thinking that tends to be both critical of Enlightenment thinking and often apocalyptic in its rejection of the development business (Escobar 1995; Rist 1997). It is at base a cultural theory:

> A cultural theory directs one to examine how the "pull of sameness and the forces of making difference" interact in specific ways under the exigencies of history and politics to produce alternative modernities at different national and cultural sites.
>
> (Gaonkar 2001: 46)

One could say, then, that the cultural move was to posit development as a modernist project (and often a failed or incomplete one at that). Development was to be construed as the cultural constitution of the social order (see Povey 1998).

Perhaps the most sophisticated account of this position has been provided by Charles Taylor (2004). Modernity for Taylor is a historically unprecedented amalgam of new practices and institutions, new ways of living and new forms of malaise. Western modernity on this view is a certain kind of social imaginary, that is to say to focus on the ways people imagine their social existence, "how they fit together with others, how things go on between them and their fellows, the expectations that are normally met and the deeper normative expectations and images that underlie these expectations" (Taylor 2004: 23). These may be self-serving and full of repressions but they are at the same time "constitutive of the real" (Taylor 2004: 183). The social imaginary has three modes of narrativity – progress, revolution, and nation – and three important forms of self-understanding, what Taylor calls cultural formations: the economy, the public sphere, and self-rule.

Taylor simultaneously paints the cultural constitution of the modern order – while all the time holding open the project of other modernities – and at the same time perfectly captures the framework within which so much of the critical work on development by the social sciences over the last 15 years has turned. Ludden's (1992) model of the India development regime, for example, is defined by (i) ruling powers who claim progress as goal, (ii) "people" whose conditions must be improved, (iii) an ideology of science to measure progress, and (iv) self-declared enlightened leaders who deploy state power for development and self-rule. Much of the cultural work has been precisely to prize apart this quarter of "configurations of power." This meshes perfectly, of course, with critical theorists of development like Timothy Mitchell (2002), Sharad Chari (2004), Tania Li (2005), or Anna Tsing (2005).

There has been no single way in which development as worlds of meaning has been explicated (see Crush 1995; Cooper and Packard 1997; Watts 2003). Some have taken the high road of development modernism and the legibility and visibility required by the state (Scott 1998); some have traced the lineages of the notion of the economy (its "invention") and the discursive construction of economic theory (Mitchell 2002; Gibson-Graham 1996). Perhaps most effective has been the ethnographic exploration of development institutions starting from Ferguson's (1994) highly generative account of how social problems are converted into technical considerations, thereby creating an anti-politics machine (normalization in Foucault's language). In his wake has appeared a raft of important studies examining, for example, the "greening" the World Bank and the deconstruction of its "literature" (Kumar 2003; Goldman 2005), forms of green governmentality and economic regulation (Drayton 1996; Agrawal 2005; Roitman 2005), the epistemic communities that help create discursively durable accounts of common development problems (Leach and Fairhead 1996), and the sorts of knowledge (conventional and subjugated) made in the name of development science (Latour 1991; Leach, Scoones, and Wynne 2005). On the other side is a body of work – generally less successful in my view – seeking to shed light on the alternative modernities under construction in Mumbai or Johannesburg (Gaonkar 2001; Simone 2005).

There is no easy way of characterizing this body of work. One part locates itself explicitly on the ground of development policy, using post-structural approaches to tease out the taken for granted assumptions of rural development programs of the operations Departments of the World Bank. Others are more concerned to paint with a larger brush on the canvas of world history. What they have in common I think is the sense in which development's many worlds of meaning can be construed as an ensemble or assemblage of "institutions, procedures, analyses and reflections," a social imaginary of "calculations and tactics that allow the exercise of . . . [a] very specific albeit complex form of power which has as its target population . . . its form of knowledge political economy" (Foucault 1991: 102). In this sense culture turns out to be a powerful weapon, not just talking truth to power but of pulling apart the "contingent lash up" (Rose 1999) that is development.

The making of a cultural revolution

> We are way ahead of the protestors.
>
> > (James Wolfenson, President of the World Bank,
> > 2000, cited in *International Herald Tribune*,
> > April 19, 2000: 19)

> [N]eoliberalism is more than an economic theory. It also constitutes
> the conditions for a radically refigured cultural politics.
>
> > (Giroux 2004: 107)

My account of the ways in which culture understood as the dialectics
of system and practice has reshaped development discourse – has
broken through the tough carapace of conventional accounts of the
state, the economy, and civil society – has for the most part operated
in an historical vacuum. Mapping cultural ideas and their institutional
points of entry is one thing, providing an account of the material
circumstances in which the cultural turn emerged and the social
forces that might go some way toward explaining the rapidity with
which culture entered development theory and practice is quite
another. The broad historical contours of such an explanation are
provided by neo-liberalism.[4] Its realities are plain to see. The world
has endured two decades and more of radical reconstruction made
in the name of a new/old capitalist orthodoxy – repeated rounds of
privatization and deregulation, tight money (for some), free trade (for
the defenseless), "adjustment programs," attacks on welfare and on
big (that is, corporate-unfriendly) government. (Ninety-five percent
of all regulatory changes during the 1990s, as inventoried by the
UN World Investment Report, were aimed at liberalizing capital
controls. The tripling of bilateral investment treaties in the first half
of the same decade was almost wholly aimed at removing "barriers"
to foreign investment.)

Suffice to say that the origins of neo-liberalism date to the 1970s,
and to the challenges confronting US economic hegemony as a result
of a crisis of over-accumulation (Brenner 2002). Faced with growing
competition from Western Europe, Japan, and East Asia, the United
States under Richard Nixon freed-up international financial markets
to "liberate the American state from succumbing to its economic
weaknesses and . . . strengthen the political power of the American
state" (Gowan 1999: 23). The institutional complex that permitted
this "gamble" of projecting US financial power outwards was the

IMF–WTO–Treasury–Wall Street nexus. Right at the heart of neo-liberalism's strategy was an assault on the state-centered development strategies of postcolonial states: markets were to be forced open, capital and financial flows freed up, state properties sold at knockdown prices, and assets devalued and transferred in crises of neo-liberalism's own making. For the Third World, the tyranny of "there is no alternative" ruled supreme; for the post-1989 socialist bloc, it was simply dubbed "Shock Therapy." What proved to be so extraordinary about the neo-liberal counter-revolution was neither its zealotry nor its bourgeois mission to create "a world after its own image," but rather its singular hyper-nationalism: that a single nation imposed its own self-description as a global norm. Its definitive statement, we can now see, was the 2002 National Security Strategy document: its absolutist modality – "full spectrum dominance." At first the neo-liberal offensive was, as Antonio Gramsci might have said, a passive revolution from above: conservative, defensive, and despotic (Silver and Arrighi 2003). Now it is coercive, military, and imperial.

What all of this amounted to was another round in the ceaseless process of primitive accumulation, what Harvey calls "accumulation by dispossession" (Harvey 2002). And the cultural turn in development, I think, must be framed by the Great Arch of modern global neo-liberalism. In this sense much of what I have described in this essay has as its reference point Karl Polanyi, who famously pointed out in *The Origins of Our Time* (1945) that markets cannot create social order, indeed they can colonize and ultimately destroy it. The market destroys the social character of three foundational but "fictitious" commodities (land, labor, and money); its corrosive effects underwrite class-based communities such as Chartism, the co-operative movement, and Owenism, each is a reaction to the destructive process of commodification (see Burawoy 2003). Re-embedding of markets, or the reactions to disembedding, produce forms of association that serve to maintain capitalist accumulation (the civil society counterweight to the anarchy of the market). Naturally, it would be too simple to posit all cultural inflections of development as the product of the giant servomechanism of market-capitalism. Robert Putnam's concept of social capital or corporate social responsibility cannot be fully reduced to market re-embedding and the violence of neo-liberalism (though neither are inexplicable without them), any more than civil society or culture is simply a "reaction" to the unregulated market. To do so would return us the bad old days of "base and superstructure" that was precisely one of the things that the cultural turn sought to redress.

So neo-liberalism and the horrors of primitive accumulation are indispensable framing devices. So too is the rise of global civil society – and it is at the intersection of these two global vectors that the culture as a "traveling theory" can be productively located. Rampant neo-liberalism – understood as a global constellation of capitalist accumulation under American hegemony rather than simply as the realm of "the economic" – and the "dynamic non-governmental system of interconnected socio-economic institutions" (2003: 9) that Keane calls global civil society, created a global forcing house within which development could be culturalized. Global civil society – Keane notes that there are 50,000 NGOs that operate "at the global level" (2003: 5) – was itself born of a complex confluences of ideas and socio-political transformations, including of course the revival of civil society in the wake of the crashing of the Prague Spring, the socio-political effects of the new virtual technologies, the deep crisis of the postcolonial state (not unrelated of course to the devastating effects of neo-liberal "adjustment") marked by the moniker "rogue" or "failed," and the panorama of conflicts emerging in the wake of the end of the Cold War. We may differ, naturally, as regards whether global civil society represents a new global public, or a coherent effort to civilize the global order, or a new cosmocratic form of politics. Global civil society is, as Keane (one of its major boosters and theoreticians) notes, a work in progress. At the very least it foregrounds the prospect of differing ways of democratic living and of rethinking what passes global markets.

To put the matter sharply, the confluence of global neo-liberalism and civil society over-determined the extent to which the dialectics of system and practice forced itself onto the development agenda. Contained within this confluence is a vastly complex set of movements and ideas; let me simply flag five:

• the global human rights movement;
• the global women's movement;
• indigeneity and the politics of recognition;
• anti-systemic movements and the crisis of the postcolony;
• the Image World of Consumer Capitalism.

Each of these properly requires a full accounting that I cannot provide here. My point is simply to confer their significance as force-fields within which worlds of meaning acquired a powerful resonance. How could such matters as patriarchy, ethnic violence, indigenous struggles, the "anti-globalization" movements (whether the revolutionary impulses of the Black Bloc or militant Islam), the volcanic eruption of locally improvised alternatives emerging from the tectonic shifts of postcolonial failure, and unmatched imperial powers of consumer capitalism be fully apprehended without the deployment of cultural categories? The extent to which intellectuals, activists, and practitioners of various stripes latched on to culture was surely shaped as much by this heady cocktail of cultural politics as by the new theoretical productions emanating from the Left Bank *ecoles*, Oxford, or Harvard.

The neo-liberal moment: culture as a form of rule

[T]he recognition of cultural domination as just as important as, and perhaps as a condition of possibility of, political and economic domination is a true "advance" in our thinking.

(Buck-Morss 2003: 103)

[P]ower often makes its presence felt through a variety of modes playing across one another. The erosion of choice, the closure of possibilities, the manipulation of outcomes, the threat of force, the assent of authority or the inviting gestures of a seductive presence . . .

(Allen 2003: 195–6)

I want to end my remarks with some reflections on the relations between the cultural and neo-liberal revolutions in development practice. To do so I want to briefly return to a period when questions of poverty, scarcity, and rule were the compelling political questions of the moment, namely the debate over the Poor Laws in Britain in which Thomas Malthus (1798) figured so centrally. The Malthusian debates the late eighteenth and early nineteenth centuries represented a debate with "popular radicalism" (McNally 1995). Paine's *Rights of Man* appeared in 1791 and it is to be recalled that it was enormously popular, selling 200,000 copies within a year. *The Rights of Man* politicized poverty – it accepted the logic of the free market but provided a program for political reform. At the heart of the popular radicalism and its opposition to political economy was a debate within the English ruling class over the paupers and what to do about them.

Malthus' ideas, whatever their class provenance, transcend demography, since his ideas were put to work at a moment of political crisis. The figure of the pauper was constituted as a social problem to be regulated, a discourse central to the shift from an Elizabethan moral economy (a paternalist system of policing to relieve and employ the poor) to a liberal mode of government of poverty fostering a space of individual autonomy and minimal state action. The "discourse of the poor," as Mitchell Dean (1991) calls it, had three new components: the emergence of a strong abolitionist strain toward the poor laws, the insistence on voluntary charity as morally superior to the compulsion inherent in the poor rates, and the substitution of make-work schemes by various contributory methods (life annuities, insurance plans, and mutual assistance). Here was a redefinition of the field of action of the state: a withdrawal of the state responsibility for relief to categories of the propertyless, and the centralization and bureaucratization of the state apparatus of relief. At stake was a liberal mode of government "distinguished by its aim of incorporating self-responsibility and familial duty within the lives of the propertyless . . . by multiplying the institutional and administrative networks which surround poverty" (Dean 1991: 18).

The liberal mode of government, scarcity as an object of regulation and discipline, and Malthus' bioeconomic laws can be productively understood as expression of modern rule, what Foucault (2000) calls the "conduct of conduct," a more or less calculated and rational set of ways of shaping conduct and of securing rule through a multiplicity of authorities and agencies in and outside of the state and at a variety of spatial levels. Foucault was concerned to show how the analytics of government constituted differing expressions of governmentality, which he articulated in terms of the pastoral (classically feudal in character), the disciplinary (the institutions of subjection from the prison to the school), and biopower (the administration of forms of life). Modern governmentality was rendered distinctive by the specific forms in which the population and the economy was administered, and specifically by a deepening of the "governmentalization of the state" (that is to say, how sovereignty comes to be articulated through the populations and the processes that constitute them).

How then do Malthus and liberal government speak to the question of scarcity and modern development? Malthus' achievement was to help construct a discourse of poverty and scarcity which challenged the radical Paine-ites and their theories of rights and to this extent it reflected both a voice of the status quo and a fear of the enfranchisement of the popular classes (of social disorder from below). In this way scarcity became central one might say to the conduct of conduct. Scarcity shifted from a pre-modern sense of periods of necessary dearth to a generalized condition of insufficiency. By the mid to late nineteenth century, after Malthus' pessimism lost some of its power, scarcity came to hold what Xenos (1989) calls "the promise of abundance." But this promise required choice, discipline, and above all a rational economic man capable of "economizing": making choices to allocate scarce resources among competing ends. But scarcity had a much broader discursive resonance. It became central to economics, which of course was significant not simply in the rise of neoclassical free market thinking but in the entire arena of economic policy and the very idea of a national economy (Mitchell, T. 2000). Scarcity became, one might say, a condition of possibility for modern rule. It represents a discursive formation precisely because it operates across many domains in and beyond the state (think of the army of family planning NGOs or famine relief agencies). Scarcity cross cuts and informs *all* of the analytics of modern government. One aspect of the cultural turn has been to shed light on how this logic operates within contemporary development theory and practice.

The relations between rule, scarcity, and poverty can be traced across the entire history of development. An important body of work – driven by the cultural turn – has shown precisely how development operates as a realm of calculation, expertise, enframing (Mitchell 2002; Li 2005). Yet the culture question plays a contradictory role in this process of global neo-liberalism. On the one hand, it has opened up the possibility for a critical assessment of the analytics of government in the developing world in the development institutions of neo-liberal rule. On the other, culture has itself become part of the arsenal of modern neo-liberal development rule. This is quite clear in the way in which community operates in relation to development projects and the debate over governance in the global South. The danger is of course that quite sophisticated senses of culture – social capital, community governance – became the basis for a sort of normalization, for the further fuelling (ironically) of the anti-politics machine. The extent to which culture in this sense is capable of securing hegemony for ruling classes of all sorts is always an open question. What are the limits then to a global neo-liberalism armed with the sophisticated (and self reflexive) deployment of culture?

One might say that this neo-liberalism has in the past few years begun to look rocky; fights within and between global regulatory institutions, resistance to free trade within the global South, and the slide into war. The ruthless underbelly of "globalization" was now exposed. America – partly in response – shifted from consent to coercion, to military neo-liberalism that in a sense compromises a certain vision of self-government. A limit to the neo-liberal project is that in large parts of the world the market-based secular nationalist project has catastrophically failed (the 2 billion dwellers of the slum world, as Mike Davis [2005] notes). It is out of this crisis of course that militant Islam, to take one example, has grown. The anti-politics machine can create the most explosive sorts of politics. And here is the rub. Some of these oppositional movements – anti anti-politics – assume cultural forms (al-Qaeda for example) that are not terribly desirable. They will require what Williams (2005) calls a "generative politics" that must inevitably be cultural but will place civil society in a dominant position by working with and through the state.

Notes

1   For useful discussions of the dialectics of culture and economy, see Crang (1997), Sayer (1997), Ray and Sayer (1999), as well as Chapter 1 and James (this volume).

2   A brief perusal of Tim Forsyth's admirable new *Encyclopaedia of International Development* (2005) reveals not only that culture has its own long entry but that development's "keywords" – its fundamental lexicon – are totally saturated with cultural thinking. Poverty – the eradication of which is the foundation stone of development – is now widely understood to be as much about "social relations" as "economic conditions." The *Encyclopaedia* properly sees poverty in terms of a panoply of needs, processes of sociocultural exclusion (gender most obviously), human rights, and capabilities.

3   The World Bank is not totally immune, of course, to such cultural analysis (this is the heart of the social capital question that I discuss here) but its center of gravity lies elsewhere.

4   This was written before the appearance of David Harvey's much fuller and sophisticated account of neoliberalism (Harvey 2005).

3 Culture and conservation in post-conflict Africa: changing attitudes and approaches

Elizabeth E. Watson

> There is no Western solution or blueprint that will solve African problems, and, unless both Africans and the West realize that, all efforts to address them will continue to fail. Peace, democracy and stability are the key for any attempt to "save" Africa, but they must be rooted in African realities and traditions. All societies, no matter how poor, have resources. The trick is to recognize them and harness them effectively . . . Africa has the potential to "take off," but it will only do so if it listens to its own beats and its own logic.
> (Manuel de Araujo, President of the Mozambican Association, 2004)

This chapter explores the way in which participatory development projects in the environmental field have recently turned towards what their proponents refer to as "traditional institutions," including "traditional leaders" or "traditional tenure systems," and attempted to use them instrumentally in their work. It argues that these features illustrate the wider shift in attitudes and approaches to culture on the part of development actors in government and non-governmental organizations, and reveal the way in which that shift is manifest on the ground. In Africa, approaches to development that are trying to build on indigenous culture are far from unusual, and they can be seen as growing in popularity because they serve a variety of different interests. For some policy-makers and campaigners searching for a postcolonial future, they represent the possibility of a return to a more African way of life, after years of colonial and modernizing regimes in which local African cultures were devalued (see epigraph p. 58). Nationalists in the post-independence period often equated forms of local culture with ethnicity, and it was their project to replace them with new, nationwide, "modern" identities. In the words of Samora Machel, Mozambican President from 1975 to 1986, "for the nation to live, the tribe must die." Missionaries, both foreign and indigenous, also had, and continue to have, a powerful influence, frequently presenting African cultures as replete with dangerous superstitions and as forms of heathenism. In recent years, as I show in this chapter, some development actors have seen culture as providing a potential refuge and form of protection from a new wave of threats to identity and to livelihoods from globalization (see also Chapter 1, this volume).

On the ground, there has been a convergence in the approach of many different development organizations including non-governmental organizations (NGOs), international development organizations, and government organizations. Many share the desire to integrate culture into development, and some, as I will show, see a focus on the community as a way to achieve this (on community, see Watts, this volume). Questions remain, however, about how to define, access, and form partnerships with the "community." Many actors see "traditional institutions" as providing a bridge to the community, and a means through which their ideals can be transformed into practice. Some have turned to traditional institutions for pragmatic and expedient reasons. They see them as representing strategic entry points into communities, through which they can achieve the highly idealistic policies of improving the environment while also empowering the community. By building on cultural institutions, it is thought that a "culturally appropriate" development can be stimulated, thus answering many of the major criticisms of development as exporting Western ideas.

The material presented here suggests that these approaches are frequently shallow and tokenistic. They tend to conflate concepts such as culture, tradition, community, and participation in ways that fail to recognize the contested nature of the institutions, practices, and sets of rights to resources. They ignore the way in which tradition is the product of a particular political history and continues to play its part in struggles for power and resources. Culture is given a prominent place in the rhetoric of development and empowerment, but development projects tend to extract and employ an idea of culture, vested in "traditional institutions," which does not necessarily correspond to the reality that exists on the ground.

Tensions and conflicts result when "traditional institutions" fail to live up to expectations or to behave as expected, and rapid U-turns in policy-making result. In the post-conflict situations, instead of providing development and improved environments, such tensions can also lead to further conflicts.

In this chapter I draw on research material from Mozambique and Ethiopia, but in order demonstrate the outcome of these projects in more depth, I focus on one case study of community-based natural resource management from Borana in the south of Ethiopia.[1] The example provides a grounded and specific illustration of the way in which culture is being objectified, valued, and integrated into development projects, and demonstrates the problematic outcomes of some of these approaches.

Community-based natural resource management in post-conflict situations

Over the last two decades there has been a proliferation of community-based natural resource management (CBNRM) projects in Africa, particularly in remoter rural regions where many of the poorest people have depended historically on natural resources for their livelihoods (Hulme and Murphree 2001). Many of these CBNRM projects seek to redress, explicitly or implicitly, some of the problems caused by processes of immanent and intentional development during the twentieth century. Although pre-colonial societies should not be seen as unchanging, or as living in harmony with nature,[2] it is evident that patterns of access to and use of natural resources have been disrupted and undermined by external factors. People have been displaced from resources that they previously used by the establishment of commercial, sometimes colonial, farms or ranches, or by wildlife parks and other conservation areas. Conflict too has taken its toll, as part of struggles for independence from colonial powers, and throughout the Cold War and post-Cold War period. There are few continent-wide studies of the extent of the conflict, but much-cited statistics suggest that "since 1970, more than 30 wars have been fought in Africa. In 1996 alone, 14 of the 53 countries of Africa were afflicted by armed conflicts, accounting for more than half of all war-related deaths worldwide and resulting in more than 8 million refugees, returnees and displaced persons" (Annan 1998). Although conflicts are continuing in many countries, such as Ivory Coast, Sudan, and Congo (where in the latter alone, 3.8 million human fatalities are estimated from 1998 to 2004),[3] other countries, including Ethiopia and Mozambique, have been experiencing periods of relative peace, and have been engaged in post-conflict reconstruction, resettlement of returned refugees, and the rehabilitation of ex-combatants.

In terms of natural resource management (NRM), post-conflict countries have been forced to cope with the way in which conflict has impacted on people's access to and use of natural resources in direct and indirect ways. The resources themselves may have been destroyed, and populations may have been displaced (and are now returning). Most insidiously, conflict may have impacted upon the social fabric and the institutions that regulate the access, use, and distribution of the benefits of those resources. Whereas in the past natural resource management projects focused mainly on the technical aspects of natural resource management, these social institutional dimensions are now understood to be vital to the sustainable use and management of resources, determining who is able to benefit from those resources, and they represent particular challenges in post-conflict situations.

CBNRM projects are participatory. They attempt to tackle some of the problems discussed above by including those who have been displaced and/or excluded by the top-down modernizing development projects of the past. CBNRM projects are also based on the premise that environmental degradation and poverty are linked; if the management of natural resources can be improved to the extent that they contribute successfully to securing livelihoods, then, it is theorized, communities will have incentives to invest in and so conserve their resources. Improving the environments will also contribute to the health and enjoyment of local people, and reduce destitution and environmentally induced migration. The participatory dimension of involving the community in decision-making and management procedures will, it is thought, ensure a sense of ownership over the development process and therefore a degree of sustainability. It will also allow the project to draw on the knowledge and skills of the local people who are thought to know best what their problems and solutions are. A great deal of work in the field of "indigenous knowledge" has demonstrated that local people frequently have rich environmental knowledge that can be integrated into a project. Finally, theories of participation argue that by involving the community, the process can contribute to its

empowerment. All of these arguments have fed into the rise of
CBNRM, which is thus conceived as a win–win–win scenario: the
environment, economic well-being, and empowerment of the
community are seen as interrelated and benefiting.

CBNRM and culture in decentralizing states

The focus of development and conservation programs on
the community exists within a broader movement towards
decentralization in many African states. Decentralization has taken
place as the resources and structures of government have been cut
back in the post-Cold War, post-socialist period, as a result of the
rise of neo-liberal market-oriented policies and the influence of
international organizations with conditional aid programs (Larson
and Ribot 2004). However, in Ethiopia and Mozambique, the post-
conflict, post-Cold War, post-socialist periods have not only been
viewed as situations in which new leaner structures of government
must be instituted and the legacy of conflict must be overcome, but
also have been presented by governments as times of opportunity in
which it is possible to bring about unprecedented empowerment for
people at the grassroots. There is also a new emphasis on the rights
of people to practice their own local cultures and religions. Whereas
governments previously viewed local practices as obstacles to
development, now they are seen as a resource that has the capacity
to promote it.

In Mozambique, the idea of community has been central to the
government's conceptualization of the newly decentralized state,
which aims to give more power to those at the grassroots. In the
post-conflict period (1992 onwards), a great deal of attention has
been paid to preparing laws through which the community can, for
the first time, register as the legally recognized manager of resources,
of land, forests, or wildlife.[4] These policies combine ideas of
generating improved tenure security to meet the needs of local
people and investors, with ideas of working with – instead of against
– customary institutions, including customary tenure systems,
customary NRM practices, and customary – or more commonly
termed "traditional" – leaders.

In Ethiopia, the decentralization policies that have been implemented
in the post-conflict period (1991 onwards) have been, if anything,

more radical. The new, post-1991 government has federalized the country using ethnicity (or nationality, as it is referred to in Ethiopia) as the organizing principle of the new federal state. The aim of this policy is to reverse the social and political hierarchies embodied in the previously highly centralized Ethiopian state. Prior to the 1974 revolution, the Ethiopian Empire was dominated by northern ethnic groups, especially the Amhara. Amharic culture was equated with the culture of the Ethiopian nation, and the religion of the north, the Ethiopian Orthodox Christian Church, was the religion of the state. Non-Amharic and non-Orthodox groups were expected to aspire to a northern Ethiopian way of life (Krylow 1994; Baxter, Hultin, and Triulzi 1996). Following the 1974 socialist revolution, the equality of different ethnic groups in the country was promoted as the official doctrine of the state, but in practice the superior position of northerners was sustained. The Ministry of Culture was active at this time but their work was very much in the style of other socialist approaches to local culture. The folklore and customs of the local people were celebrated, but situated in the context of a modernizing revolutionary socialism that viewed local cultural institutions with distrust, and which tended to represent forms of local culture as "backward" and potentially oppressive (Donham 1999).

By contrast, in the post-1991 period "Every Nation, Nationality and People in Ethiopia has an unconditional right to self-determination, including the right to secession" according to the 1995 Constitution (Article 39.1 1995: 13).[5] A "nation, nationality or people" is defined in the constitution as "a group of people who have or share a large measure of a common culture or similar customs, mutual intelligibility of language, belief in a common or related identities, a common psychological make-up and who inhabit an identifiable, predominantly contiguous territory" (Article 39.5 1995: 13). Each "nation, nationality or people" now has the constitutional right to its own local governing institutions and to make its own policy decisions. This decentralization policy is thought, like participatory development programs, to enable policy decisions to be more effective, locally appropriate, and empowering, as they can build on local knowledge and involve people directly. They have also been accompanied by explicit policies aimed at promoting culture. The preamble of the 1995 constitution refers to "our rich and proud cultural heritages" (1995: 1), and Article 39.2 states that "Every Nation, Nationality and People in Ethiopia has the right to speak, to write and to develop its own language; to express, to develop and to promote its culture; and to preserve its history" (1995: 13).

The decentralized states' approach to local communities and their culture is an important key to understanding the way in which development project planners have started to see local cultural institutions, not as something to be overcome, but something that can be legitimately harnessed for CBNRM. For these states, the emphasis on culture and community provides positive meaning to the structure of the decentralized state by portraying decentralization as a kind of second wave of decolonization. It contends that the countries were not liberated after the dissolution of the empires of which they were a part (Portuguese and Imperial Ethiopia respectively). Both of these imperial regimes were replaced by top-down modernizing socialist regimes, which also attempted to impose another set of values and a particular order. The new emphasis on the community and on culture is no doubt in part influenced by global discourses of multiculturalism, but is also part of the post-1990s governments' attempts to distance themselves from previous regimes and present their own as having a fresh and attractive approach and rationale.

Community and "traditional institutions"

In "the field," the turn towards forming partnerships with "traditional institutions" for CBNRM has emerged in part because of innate difficulties with implementing community-based projects. The rhetoric of CBNRM implies that the community is homogeneous, that it knows what it would like, and what would be in its interest. It implies that the community is easy to identify. On the ground, though, all communities are made up of socially differentiated interest groups, and they are not always clearly bounded or easy to locate and demarcate (Leach, Mearns, and Scoones 1997; Guijt and Shah 1998; Agrawal and Gibson 1999). In Mozambique, for example, one government development worker explained in an interview, "[when we started] we wanted to work with the community, but when we went there, we found there was no community, the people were living all over the place" (Author's interview, Manica Province, 1999). Given the heterogeneous nature and blurred boundaries of the communities, many development workers have employed "traditional institutions" as the signifiers of community.

"Traditional institutions" are portrayed in some development literature as ready-made grassroots institutions. They are also thought to be repositories of indigenous environmental knowledge, and, as their decision-making responsibilities extend to matters concerning forests, land or water, they are thought to be key institutions for NRM. When they are given the label "traditional," they obtain what Weber described as the legitimacy of the "eternal yesterday" (Weber 1978). In Africa where written records are particularly poor, they are considered to have been in existence, unchanging, from "time immemorial," and, therefore, to be authentic representatives of their communities. Although it may not always be made explicit, it is generally assumed that, through "traditional leaders," the multiple aims of improved environmental management, economic development, and empowerment can be achieved. In the post-conflict situation they are thought to bring a valuable sense of continuity, of social cohesion and social security: they can therefore help a community recover from the conflict experience.

For example, in Mozambique, the United Nations Food and Agriculture Organization, FAO, was heavily involved in developing the land, forestry, and wildlife laws that recognized the communities as legitimate resource holders, decision-makers, and managers. It was also involved in programs that trained local communities to use Global Positioning Systems to delineate and register community resources. Excerpts from a 1996 FAO news report about community-based projects in Mozambique in their early stages (see Box 3.1) exemplify in a classic way how African agricultural communities are repeatedly portrayed in development literature, "struggling" from "dawn to dusk" with only a "simple hoe." The excerpts also highlight the expectations placed on these community projects: they are seen as enabling communities to overcome the disruption caused by war, and to protect them from being further displaced by the expansion of commercial farming. It also shows the way in which working with the community is translated into supporting "traditional land tenure systems," and the involvement of "traditional leaders." From interviews with those involved in FAO-backed projects, it was evident that community resources were understood by them in terms of the territories under the jurisdiction of a local traditional chief, known in Mozambique as a *regulo*. Those involved in mapping community lands, in order to delimit, demarcate, and certify community rights, were engaged in the difficult task of establishing the boundaries between the territories of neighboring *regulos*.

Box 3.1

Excerpt from "News and Highlights," United Nations Food and Agriculture Organization

New land law entitles people to security in Mozambique[6]

A simple hoe is the only tool with which Elisa Matlombe keeps her five children alive. From dawn to dusk, she struggles on her tiny patch of land in the southeast corner of Mozambique, growing a little maize and groundnuts.

Nor is there a Mr Matlombe to bring money home from the mines in neighbouring South Africa or a job in nearby Maputo, the capital of Mozambique. Her husband was beaten to death by Renamo insurgents during the country's 16-year war. As if this was not enough to endure, Elisa faces the constant threat of eviction. She has no clear title to the land and a local developer has laid claim to it.

But help may be on its way for Mrs Matlombe and the hundreds of thousands like her who fled from their fields during the war, if the Mozambican Parliament passes a new land law drafted with assistance from an FAO project.

The new law, taken up by Parliament in November, would give legal backing to a land policy designed to recognize both current realities and traditional land tenure systems . . .

"We came up with a new concept called a rural community, which would be defined by customary occupation," said the FAO team leader. "This covers a bigger area than what they're using at the moment. It includes areas in fallow or which are still uncultivated, forests, water rights, grazing land, areas which are being held for future generations or those which are sacred or culturally important. Within these areas, the local community is free to allocate, inherit or transmit land among its members according to local practices . . ."

While the draft law still allows investors to gain access to community land, this can no longer be done behind the community's back. "The new law gives rights to traditional leaders, and no land can be ceded to developers without consulting first with the community," said João Muthombene, coordinator of ORAM, an organization that defends farmers' rights in land disputes.

From the World Bank, Shelton Davis has written how there have been increasing attempts globally over the last two decades to bring culture into the development arena. He cites Mozambique as one of the earliest countries to take up the challenge of integrating culture and development, with a "National Conference on Culture" in Maputo in 1993. The culture and development "strategies" are summarized as focusing on the fields of "material culture, expressive culture, intercultural communication, and the protection of cultural and natural heritage" (Davis 1999: 29). Its aim is to "buffer some of the turbulent effects of the globalization process, strengthen community pride and social cohesion at the local and national levels, and contribute to poverty reduction and sustainable development through the fostering of more peaceful societies" (Davis 1999: 29). These broader national and international policy contexts have influenced the way in which community-based projects have been developed, but, in the field of CBNRM, culture has been interpreted much more narrowly and instrumentally. There is little discussion of the arts or of the exploration of cultural meanings here. Instead, the aim is to identify the cultural institutions, the "traditional leaders" and forms of "customary tenure" that "work," and build partnerships with them in order to achieve particular ends; that is to improve NRM and reduce poverty (compare Elyachar 2002). In the next section, I explore one case of CBNRM from Ethiopia to illustrate some of the problems that can arise.

CBNRM and "traditional institutions" in Borana, Southern Ethiopia

An analysis of the CBNRM projects attempted in Borana since the late 1990s can reveal in detail how policy attitudes towards culture have changed in recent years. The projects show the specific ways that culture, in the form of "traditional institutions," are thought to be integrated into development projects as a way of mobilizing the community and improving environmental management.

Borana Zone in Ethiopia is situated in the region known as Oromiya because it is the area inhabited by those of the broader ethnic family group known as Oromo (see Figure 3.1). These people share the same language, *Afaan Oromoo*, and also have a belief in a shared history, together with some similar customs and institutions. The Borana people inhabit the lowlands on the Ethiopia–Kenya border;[7]

the lowland Borana region is an area of grasslands with pockets of bush and forest used mainly for pastoralism, although grazing is limited by the availability of water for animals. There are different sources of water that vary throughout the year: deep wells; rain-fed ponds and reservoirs; seasonal surface water and rivers; and boreholes. Most important are nine deep well complexes, known as *tulaani saglaani*, that contain water throughout the year. They are as old as, if not older than, the Borana residence in the region, and they are of great ritual, as well as practical and material, significance to the Borana (Helland 1997). The Borana are mainly cattle-keeping pastoralists, although they combine this with the rearing of small stock and, increasingly, with the rearing of camels. They think of themselves as pastoralists first and foremost, but agriculture is becoming more common.

The case of the Borana illustrates the way in which any recent promotion of culture represents a reversal of previous attitudes. The culture of the Oromo-speaking groups was demonized in Ethiopian

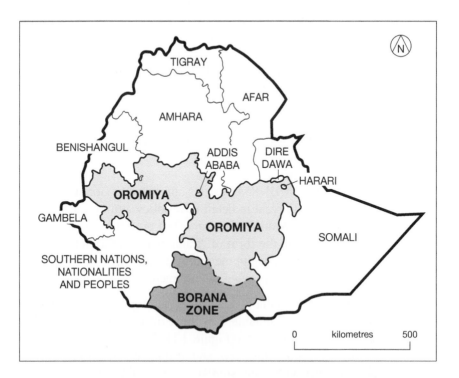

Figure 3.1 Map of Ethiopia showing location of Regional States and approximate location of Borana Zone.

historiography, which reflected the particular dynamic of the dominant views and power structures of the Ethiopian state and its personnel in the twentieth century:

> In much Ethiopian historiography the historical migration and expansion of Oromo-speaking groups has been imagined as the culmination of a series of barbaric – that is Cushitic and Muslim or "pagan" – invasions of an ancient Semitic-speaking, Christian or Judeo-Christian, monarchical, literate civilization in the heart of darkness.
>
> (Hultin 1996: 83–4)

The people of the Ethiopian borderlands (including the Borana), who were incorporated into the Ethiopian Empire by Emperor Menelik II at the end of the nineteenth century, tended to be referred to by the pejorative appellation, *shanqilla*. This term had racial connotations, meaning people who were black and physiologically of a more African appearance than the lighter-skinned northerners. The term was also used to mean slave.[8] Through the use of cultural signifiers such as this, the people of the margins in general, and their cultural practices in particular, were constructed as inferior to those of their northern counterparts.[9] In this context, the attempts of CBNRM projects to work with culture in the form of its traditional institutions represents a radical departure, and has been seen as having the capacity to reverse many years of prejudice.

In Borana, at the end of the 1990s, there was broad-ranging agreement among both government and non-government organizations, that previous development projects had not only failed, but that they had actually undermined livelihoods and caused environmental degradation in the area because they did not take into account the institutional dimensions of NRM. Governmental, World Bank, and NGO organizations had built reservoirs and ponds, and sunk boreholes in the area, with the aim of improving the productivity of the region with little success. Little or no attention was paid to regulating access to the grazing lands surrounding the new water points, resulting in an over-concentration of people and animals around water points, and environmental degradation.

In contrast to this picture of environmental destruction, the Borana rangeland management system has been recognized in the international literature as one of the most productive in the world (Scoones 1995: 12). Thus, in the 1990s, many development organizations in the region started to argue that the environment had been degraded as a direct result of the way in which the Borana traditional institutions were overlooked and undermined under previous prejudicial and modernizing regimes. A report of SOS Sahel, a UK NGO, explains: "[A] fundamental misunderstanding of pastoral land management, and in particular pastoral tenure systems, has undermined traditional institutions and the environment for which they were once responsible" (Boku Tache and Irwin 2003: 5). A recent review of the Borana situation by GTZ/BLPDP (a collaborative project between the German government organization, GTZ, and local Oromiya government) explains that the "core problem" in the region is that the establishment of modern government structures has "weakened indigenous decision-making structures in NRM" and caused "inappropriate utilization of rangeland," "reduced livestock condition," and, ultimately, the "death of people" (author's interview with GTZ/BLPDP 1999; see Watson 2003).

From the end of the 1990s, the literature and attention of development organizations in the area started to argue that if the "traditional" Borana practices and institutions that had managed the environment so well for so long could be understood, located, and integrated into development projects, then successful CBNRM could be achieved. Improved NRM and improved environments were seen as within reach by drawing on – or returning to – the traditional Borana way of doing things. The Project Manager of Save the Children US, explained:

> The failure of water development projects in the past was due to big ponds which were dug by machine. The capacity of the ponds was too big to be managed and maintained by local people, and the traditional knowledge was not taken into account. Instead, to reverse this trend we identified community water managers. The SCF-US team consulted the community water managers and decided that a Resource Management Committee should be elected. The people elected were often knowledgeable elders and indigenous leaders. Most of the people chosen were people close to the traditional leaders, if not the traditional leaders themselves.
>
> (Author's interview, 1999)

Most prominent and important of the Borana "traditional institutions" is the *gada* generation system that organizes all adult men into particular generation-grades, each with certain responsibilities to the wider society. The generation-grades are led, administratively and ritually, by a council of men, who remain in office for a period of eight years, and who themselves are led by one appointee known as the *aba gada*, or "father of the *gada*." Halfway through this council's term of office, men from all over Borana come together in a meeting known as the "meeting of the multitude," at which those who will serve in the next council are elected. Matters of concern to the Borana people including security issues, practical environmental issues (for example, the use of and access to land and wells), and ritual issues are all discussed at this meeting. A set of principles known as the *aada seera* that summarize the principles delineating expected behavior in relation to everything from Borana style of dress, to land and water use, to interethnic relations, are also discussed, amended or ratified at the meeting; although largely unwritten, they have a relatively formal status. At a more local level, there are also men who hold positions that relate more directly to the access to and use of particular water wells or who are responsible for a particular group of households, and their movements. Again, these decision-makers are expected to act in accordance with the *aada seera*.

Many development organizations cited the *gada* system when they were explaining how they worked with the community in their CBNRM projects. For SOS Sahel, stress was placed on the way in which the *gada* and other institutions were participatory and democratic. They therefore were seen as compatible with current development priorities and values: "The Borana society achieves consensus on important community issues through open and participatory systems of management" (Huqqe Garse 1999). In a later publication, SOS Sahel go further and emphasize the way in which the *gada* is a pan-Oromo institution, that is, a cultural institution shared by all Oromo-speaking groups. Such a framing places the CBNRM initiatives within the context of an Oromo-wide identity and cultural consciousness, shared by all people in Oromiya (see Figure 3.1):

> The *Gadaa* is a pan-Oromo organization. It is an enduring institution that has been in operation for at least the last 600 years. Long historical and political processes in Ethiopia have weakened it in the central parts of Oromiya. However, in spite of all this, the Borana *Gadaa* remains one of the most intact traditional institutions in Ethiopia today ... The project has therefore started to work with the *Gadaa* as an entry point for the representation of local resource users and interest groups, who have been omitted from modern Government management structures and systems.
>
> (Boku Tache and Irwin 2003: 33–4)

On the ground, these approaches were employed through meetings with traditional leaders such as *gada* council members and the *aba gada*.[10] Attempts were made, as in the SCF US case, to use these meetings to set up CBNRM committees. In addition, the existing structures such as the *gada* could also be drawn upon to provide decisions on certain environmental matters, and to advise the community after discussions with these state and non-state organizations.

These approaches, however, have failed to live up to the expectations that were placed upon them. Within a year of starting their program of working with traditional institutions, by 2000, GTZ/BLPDP had already abandoned it, commenting that the "traditional system" had been undermined by the processes of history, and that it was no longer possible to go "back" to it:

> Now the traditional situation is not well-functioning. In all places the *aba dheeda*[11] has been replaced by the PA Chairperson [head of the lowest rung of administration in the post-1974 state]. We don't try to bring back the traditional system. If the community thinks the traditional system is functioning then well and good, but if not then no. It is difficult to bring back the traditional system. Things have already turned 180° – it doesn't make sense to try and turn it back again.
> (Author's interview with GTZ/BLPDP 2000)

My discussions with local people about attempts to work with local cultural or "traditional" institutions revealed that, far from being experienced as empowering as the rhetoric suggests, the initiatives have been viewed with intense suspicion. In the next section I examine the reasons why these approaches have proved problematic from the perspectives of the majority of development organizations in the region and of the local people themselves.

Politics of culture

There are several reasons for problems with the traditional institutions' approach to development. In general, they are connected to the way in which, in order for organizations to argue that these previously overlooked institutions are valuable and can have a role in development, their roles are exaggerated and they are idealized. Working with "traditional institutions" is presented as "going back" to a condition that existed before the establishment of the modern state, and all that the state brought with it (see Watts, this volume). Culture, tradition, participation, and empowerment are linked in a way that conjures up images of an autonomous people living in harmony with each other and their environment, in a world mediated by custom and culture. This rhetoric not only fails to acknowledge the political contestation that is contained within each of these terms, but its feel-good language can also be seen to disguise some of the political processes taking place. When these traditional institutions are viewed in their historical context, and in their wider geographical context, which acknowledges that there are other ethnic groups in the region, a very different and more political picture of these institutions emerges.

In the area in question, the access to and use of natural resources has in fact been dominated historically by competition and conflict between the Borana and their neighboring ethnic groups (Helland 1996; Gufu Oba 1996; Bassi 1997). Other people in the area include the Gabbra, who are a minority Oromo camel-keeping group, and various Somali-speaking groups who are also specialist camel-keepers, including the Garri people. The Borana have controlled the majority of the territory and the most precious water sources since at least the seventeenth century. The Gabbra, Garri, and some other Somali groups have been able to use water, forest, and grasslands in return for paying ritual tribute to the Borana sacred leader, *qallu*, and thus demonstrating respect and allegiance to the Borana (Gufu Oba 1996). Despite this, violent conflict has flared up between these groups from time to time and has worsened in recent decades.[12]

The conflict is frequently and popularly understood as being based on historical and tribal enmity. This is worth noting, as it demonstrates that although culture in the form of "traditional institutions" is now thought to have the capacity to help communities recover from conflict, culture has often been blamed for that conflict (in the past, but also today). The term "ritualised killer complex," coined by the anthropologist Schlee (1994: 135) to describe the way in which killing an enemy in Borana culture had ritual and social significance, has been extracted from the context in which it was written and used indiscriminately by politicians as a "sound bite" to explain the conflict (Gufu Oba 1996; Baxter, Hultin, and Triulzi 1996). The Somali are also frequently portrayed as people who are unpredictable and in whose culture violence plays a central part. For example, Besteman's study explores the similarity between different accounts of Somali conflict: "the predominant explanation included claims about Somali history ('ancient rivalries'), culture ('traditional weapons' used while Somalis 'raided,' 'warred,' and 'roamed the arid plains'), and politics ('feudal fiefdoms,' 'autonomous clan states')" (Besteman 1996: 122).

To view the conflict between the Borana and the Somali as purely a cultural matter is to ignore the *political* processes at local, national, and international scales that have shaped it. The region is often thought of as remote and as having escaped from the impact of foreign colonialism or international politics, but the Borana and the Somali groups have both had particular relations to the Ethiopian and Somali states respectively. These have impacted on their relationship with each other, and on the way in which they relate to the natural resources in the region. This political history has proved crucial to the way in which CBNRM projects have unfolded.

The history of the impacts of the modern states in the region is, in fact, a story of three things: the progressive drawing of lines around fixed areas of natural resources, exclusive allocation of access to those natural resources to particular groups often defined by ethnicity, and the contestation of those definitions. Bassi (1997) describes, for example, how the British boundary commission, drawing up the boundary between Ethiopia and Kenya, divided the land used by Borana and other people into two, but then designated the land on the Ethiopian side as Borana, and on the Kenyan side as Gabbra and Garri. In the 1930s, the Italians recruited Somali peoples to help them in their invasion of Ethiopia, and the Somali groups used their access to arms to consolidate their access to and control over land and water resources (Helland 1998). In response, the Ethiopia–British alliance recruited and armed Borana to oppose them. In the 1960s and 1970s, a similar power struggle was played out, as, in a prelude to the Ogaden war of 1976–8, the Somali government supported a Somali irredentist movement in Southeast Ethiopia. Again, the Ethiopian Empire armed Borana to resist them. Thus the relations between Borana and Somali groups became entangled with international and Cold War politics, the competition for territory, and the contestation of the Ethiopia–Somalia border.

From the time when northern Ethiopia incorporated Borana into its Empire at the end of the nineteenth century, the ritual leaders, *qallu*, were made *balabbats*, intermediaries between the Borana people and the Empire. Their responsibility was to collect and provide tribute to the state. Borana people also had to give labor tribute to the soldiers and administrators of the Empire who were located there. This rather loose form of control and extraction was made much firmer under the *Derg* regime that followed the revolution in 1974. Under this government, the *madda*, described by Helland (1996: 144) as "a vaguely defined [Borana] unit with a permanent water source, such as a well complex at its centre" was used to designate the area and to describe people living within Peasant Associations. These were the lowest level units of the revolutionary state, through which the "development and co-operation" of people, as well as their taxation, were organized:

> Peasant associations, as agents of the state . . . played a much more active role and interfered much more in local affairs than had been the case with *balabbat*. Initiatives were probably generated by a need to maintain control over possible opposition and subversive activities directed against the state, but occasionally they had more direct effects in the local context. For example, attempts were made to take over the management and control over the water, to require local pastoralists to apply for permission to move from one association to the next, or to use the judicial committee of the peasant association to resolve conflicts. Increasingly, the peasant association also became the main instrument to extract resources form the Borana, first in the form of taxes and various membership fees, later "voluntary" contributions and delivery quotas of livestock at state determined prices, as well as recruits for military service.
>
> (Helland 1996: 145)

The geography of the Borana, like that of many other mobile pastoralists, is not structured into bounded and fixed spatial areas. As movement is important to seek pasture and water, their mental maps are structured around focal places and paths (Ingold 1986; Schlee 1992). Thus using institutions such as the *madda* as a bounded institution, involves a fundamental change in the nature of that institution. Helland, for example, stresses that the *madda* refers only to "all the people and all the animals who use the wells, in addition to all range resources served by the well complex . . . There is no evidence, however, that the people who happen to reside within this unit are organized in any way with reference to the territorial unit" (1996: 144). The *madda* was used to divide land into particular territories and the people into fixed groups within them. This established firm boundaries that may not have existed before, limited the movement of the people, and furthered the state's control.

The period that followed the end of the *Derg* regime, with its emphasis on decentralization, self-determination, and cultural freedom, might therefore have been expected to represent a positive change for the Borana. Unfortunately this has not proved to be the case. This period, like those before it, has been dominated by the process of drawing lines across the land, and the delineation of territories, but the process has become, if anything, more fraught. The establishment of the new ethnically federalized state has required a boundary to be drawn across an area where Somali and Borana used to share water and grazing lands, to some extent. The area on one side of this line has been designated Oromiya, and exclusively for the Oromo (including Borana) people. The area on the other side has been designated Somali. Such a designation is problematic enough in itself, but it has been accompanied by two processes that have exacerbated the situation.

The first process relates to refugee resettlement programs in the region. From the 1970s to the 1990s, the United Nations High Commissioner for Refugees (UNHCR) has been involved in resettling refugees from conflict into the Borana region discussed here. Most of these refugees are Gabbra, Garri, and other Somali people. According to Bassi, the "UNHCR registration and repatriation procedures were based simply on each individual's own statement of identity" and "the incentive of the repatriation grant [and further future assistance to refugee communities] no doubt opened the way for the 'return' of people not originally belonging to Borana province" (Bassi 1997: 39–40). In the post-1991 period of ethnic decentralization, Somali and Oromiya regions were designated in terms of the ethnicity of people in the region. Areas with large numbers of Somali peoples became Somalia; where large numbers of Borana, Oromo. Many Borana people claim that the actions of UNHCR and other development organizations have led certain areas, and the valuable water points within them, to be designated Somali Region, because of the large numbers of returned refugees. As Bassi summarizes: "by transferring as many Garri-Gabbra 'returnees' as possible into the administrative units of Borana zone, the old goal of getting access to Borana-controlled resources may, in fact, be achieved" (1997: 40). This situation erupted into violent conflict throughout the 1990s, and in 1999–2000 when my research was being carried out.

The second process relates to the way in which the relations between the Borana and the government have worsened in the post-1991 period as the Borana have been viewed as a potential source of support and protection for the Oromo Liberation Front (OLF). This movement declared war on the post-1991 government, whom they consider to be dominated by northern groups (particularly the Tigre). This, they argue, continues the subordination and exploitation of southern Ethiopian groups in general, and the Oromo in particular. The Oromo groups together form the single largest ethnic group in Ethiopia, and the OLF thus presents itself as a significant threat. Although the OLF "had not, however, established themselves in any significant way among the Borana and their pan-Oromo ideology had not made much of an impact on Boranaland" (Shongolo 1996: 267), the perception remains that the government views the Borana with distrust because of the Oromo connection. The Borana are symbolically important to the Oromo ethno-nationalist movement, as they have the most "intact" Oromo traditional institutions, and are considered by many to represent their cultural homeland.

These two processes together have meant that during the post-1991 period, rightly or wrongly, there is a perception among the Borana that the government is pro-Somali and anti-Borana. It is in this context that the CBNRM projects in Borana have failed to live up to the high expectations that its proponents held for them, and it is not surprising that local people view them sceptically. Many of the organizations discussed above (such as GTZ) work in direct partnership with the state. Other NGOs find it difficult to work independently from the state, and it seems unlikely that local people perceive them as independent. Two of the most important principles identified by theorists for CBNRM in areas where resources are shared are the necessity for a clear definition of the boundaries of the natural resources and for a clear definition of the membership of the group who use them (Ostrom 1990; Agarwal 2001). Thus, the first stages of a CBNRM involve a decision regarding the unit which will be used for natural resource management, a decision on how the boundaries around those resources should be defined, and a decision on who comprise the legitimate community of users.

This exercise to some extent mirrors some of the processes that have gone before, such as the establishment of the Peasant Associations, which were set up in the name of development but were a prelude to the control of people and the extraction of resources. Moreover, the exercise is reminiscent of, and draws immediate attention to, the way in which the contested boundary between Oromiya and Somali Regions has been drawn, and areas allocated to the Oromo or Somali respectively. In this context, the CBNRM activities are viewed as similarly political, as they too have the potential to be part of the ongoing struggle of different groups to extend their control over resources. When discussions are held with "traditional leaders" about Borana life and culture, they take place in a context in which the culture of the Borana has been objectified and appropriated by the OLF. Such discussions and attempts to form partnerships are seen, therefore, as being full of political intent. They are not viewed as being concerned with genuine empowerment or culturally appropriate development, but as motivated by the desire to monitor the Borana more closely, because of the suspicion that the Borana may be harbouring ethno-nationalists. Alternatively they are viewed as attempts by the state to capture and therefore neutralize the cultural capital of Borana institutions that is currently used to the advantage of the ethno-nationalists. The central principle of these CBNRM projects involving the partnership between external agents (including the state) and the traditional institutions is undermined by distrust, sometimes on both sides. One man's words reflect this suspicion of the government's intentions:

> The principles of the Borana are now becoming weaker. They are becoming weak because the government started interfering. They started to attend the "meeting of the multitude." They are putting pressure on the administrators of the wells, who are no longer permitted to punish those who do wrong. The system is still there but it is weaker . . . The bad thing is that the whole thing is becoming political . . . The hand of government is moving more and more into Borana and into the life of the Borana. People who used to live in peace together are being put against each other by the government. They plant fear in your mind, and then you don't speak freely. Even you can't speak without peering through the chink of light in the door to see who's listening. Most difficulties come from the government – there are others, but most come from this.
>
> (Author's interview, Man B, 2000)[13]

In this context it is not surprising that GTZ/BLDPD backtracked in their attempt to work with "traditional institutions." They found the "traditional institutions" did not constitute ready-made institutions that could be harnessed easily for improved NRM. They found that the institutions lacked a straightforward, bounded, spatial dimension, and that they were embedded in complex social and political processes that made them difficult to use.

SOS Sahel are continuing in their approach, as they do not see politics as having the potential to corrupt Borana institutions as the Borana man cited above appears to believe. There is a danger, acknowledged in my interviews in 2000, that in taking this approach they might be seen as pro-Borana and pro-Oromo; such a position would be highly problematic in this highly charged political situation. This danger is not ameliorated by some SOS Sahel literature, however, which makes scant references to non-Borana or non-Oromo people. In a 2003 summary, for example, non-Borana people are described as "returnees," relegating them to the status of refugees and implying that their rights to use resources are secondary. Although this may be the case for some people, the complex history of rights to resources and the way it has been contested over time shows that the claims of different groups need further exploration and sensitivity.

Conclusion

CBNRM projects in Borana turned to traditional institutions in the hope that they would provide a ready-made set of institutions that could be harnessed for more culturally appropriate development. The traditional institutions were seen as the legitimate representatives of the community, who could be used to define and to access that community. The traditional institutions had also historically been involved in the management of trees, land, and water resources, as well as in resolving conflict between people and groups. In the post-conflict situation, at a time of social disruption and environmental degradation because of ill-conceived development projects, it was therefore hoped that partnerships could be formed with traditional institutions in order to achieve the multiple aims of social cohesion, improved livelihoods and improved environments. At a juncture when other approaches had failed, working with traditional institutions could be presented as a fresh approach. When many organizations needed to reach people stretched out over large areas, this approach built on existing institutions and knowledges.

Although many of the development organizations active in Ethiopia acknowledged the cultural histories of the areas where they were working, the traditional institutions were rarely seen as being implicated in these political processes. History was seen as something that happened to institutions, undermining their roles and responsibilities without involving them or changing them fundamentally. If traditional institutions are conceptualized as outside political processes, it is easier to think of forming partnerships with them and to strengthen them to facilitate development and create an improved environment. These development programs are themselves thought necessarily to be outside the realm of politics.

Instead, the research has shown how historical-geographical processes unfolded in the region, so that the Borana, in common with other groups in the area, became involved in political struggles of various kinds. Assertions of identity and cultural practices were and are intertwined with struggles for natural resources. Claims to natural resources are fought in terms of claims to be the "traditional" inhabitants of an area, claims to the ritual tribute of other groups who wish to use those resources, and claims to have the "correct" set of rules to use those resources. Over time these struggles have become interwoven with national and international political processes, particularly the division of territories into states and regions. Identity and claims to access have become equivalent to each other and they are both political.

CBNRM projects have experienced difficulties because they are inherently and inevitably political. The process of identifying which community to work with invariably means promoting one community over another.[14] Recognizing traditional institutions as the signifier and legitimate representative of a community is likely to mean promoting one ethnic group or clan over another. In a context where historically rights to resources are defined in terms of ethnicity, this possibility is heightened dramatically. In the Borana case, there had been some degree of resource sharing in the past. Recognizing only the Borana traditional institutions has the capacity to entrench their dominance in the region and produce an increasingly exclusive control over resources. This process is exacerbated by the way in which CBNRM projects delineate specific areas of natural resources over which traditional leaders are thought to have jurisdiction.

A further reason why the CBNRM projects have proved problematic was the legacy of the distrust between the Borana and the government, a distrust that also extended to other development organizations working in the region. The political power plays that are part of the CBNRM processes are likely to increase these feelings of distrust. The Borana believe that the government and others are carrying out CBNRM projects due to political and security concerns, rather than philanthropic motives. The context of political struggle and mistrust means that there is little room for genuine partnership. Discussions about development goals and priorities are constrained by a focus on improved natural resource management, and by ideas that this can only be achieved by delineating certain areas and defined users. In this context, the local cultural relations that people maintained with natural resources take second place to these particular conceptions of natural resource management and conservation.

In CBNRM projects, therefore, culture is interpreted in a narrow sense as a resource that can be used to achieve certain aims. The more creative open-ended idea of culture as "structures of feeling" (Williams 1977; also see Chapters 1 and 10, this volume) appears only in this case as a force of nationalism which has the potential to result in conflict. The development organizations are perhaps therefore right to be afraid of it, but if they ignore it, and portray some idealized notion of culture and traditional institutions then they will experience other problems. The projects that they promote become intentionally or unintentionally part of the very political struggles that CBNRM projects were designed to circumvent. Further breakdowns in trust and more suspicion may result, leading not to a culturally appropriate development, but to heightened animosities.

Notes

1   The fieldwork on which this chapter is based was carried out as part of research project entitled "Reconstruction of Renewable Natural Resource (RNR) Management Institutions in Post-Conflict Countries," or "Marena project," funded by the Department for International Development, UK. Fieldwork was carried out in Mozambique in 1999, and in Ethiopia in 1999 and 2000. Each of the three visits was for between one and two months. The paper also draws on findings from fieldwork in Mozambique by colleagues Black, Serra, and Schafer. A more detailed and direct comparison of the Mozambican and Ethiopian post-conflict situations can be found in Black and Watson (forthcoming), which also discusses some similar elements of CBNRM projects in both countries. Mozambique and Ethiopia are two countries in East Africa which are rarely compared, but which experienced similar periods of civil conflict under socialist regimes from the 1970s to early 1990s. They have also both experienced periods of relative peace since the early 1990s.

2   See, for example, Kopytoff (1987) or Spear and Waller (1993) for accounts of the high levels of change and mobility among pre-colonial societies.

3 Figures from the International Rescue Committee (IRC) on http: //www.theirc. org/index.cfm/wwwID/2132 (accessed May 2005).

4 A new land law was implemented in 1997 and amended in 1998 and 1999. A new forest and wildlife law was put forward in 1999 and approved in 2002.

5 This is referred to informally as the right to divorce. A nationality (defined by shared ethnicity) has the right to divorce if "a demand for secession has been approved by a two-thirds majority of the members of the Legislative Council of the Nation, Nationality or People concerned" and certain other conditions are met (Article 39.4a, 1995: 14).

6 Report by Leyla Alyanak (http: //www.fao.org/News/1996/961204-E.HTM, accessed, April 2005).

7 The Borana in fact straddle this border, although they are the dominant ethnic group in the region on the Ethiopian side, and in a minority on the Kenyan side. Although Borana people may move across the border (freely in search of pasture or in response to political crises), many remain on either the Ethiopian or Kenyan sides. The policies with regard to CBNRM and governance more generally on either side of the border have been quite discrete. This paper focuses on the policies and experiences on the Ethiopian side.

8 On the blurring of racial and cultural meanings and categories, see Chapter 1 (this volume).

9 See Donham and James (1986) for more on the history of people of the Ethiopian borderlands.

10 See Watson (2003) for excerpts of the discussion in one such meeting.

11 The *aba dheeda* is an individual customarily responsible for access to an area of grassland (known as *dheeda*).

12 Although the conflict was mainly between the Borana Oromo and Somali speaking groups, there had also been some conflict between Borana and Gabbra. The Gabbra people are viewed with distrust by some Borana because of what are viewed as their Somali traits: although they speak *Afaan Oromoo* like the Borana, they keep camels, make houses, and to some extent also dress more like Somali peoples. The label "post-conflict" only fits the Ethiopian situation in the sense that it refers to the situation and sets of policies that followed the end of the large-scale civil conflict in 1991. There has been localized conflict since then, and also a war with Eritrea, but neither have been on the scale or had the widespread impact of the previous protracted conflicts.

13 Interview translated *in situ* from *Afaan Oromoo* by Adi Huka.

14 The process of recognizing one leader over another also may provide one leader with political recognition and power, while denying recognition and power to others. The process of recognizing a particular leader also assumes that they represent the community, but the act may strengthen their power and position over others in the community. This chapter cannot explore these processes, but some discussion can be found in Black and Watson (2006).

4 ▶ Indigenous groups, culturally appropriate development, and the socio-spatial fix of Andean development

Sarah A. Radcliffe and Nina Laurie

Latin America's diverse and numerous indigenous peoples have historically had a problematic relationship with development, due to their subordination as cheap labor in market and non-market economic relations, and their marginalization from mainstream development interventions.[1] For the region's estimated 40 million indigenous peoples,[2] the experience of development has been one of racism and the predominance of non-Indian priorities, as development planners and states have largely seen Indians and indigenous cultures as an obstacle to modern and Westernizing development projects (Stavenhagen 1996).[3] In this context, indigenous populations contest the cultural economies of which they form an integral part in order to rework unequal relations, resist changes that worsen their position, and become resilient in the face of discrimination (Pacari 1996; Katz 2004). In the words of a report on Latin America, "indigenous people demand an equal position with other citizens and full access to the results of development . . . [but] they do not want to be developed necessarily within the framework of dominant development models" (Pirttijarvi 1999: 10–11).

As a consequence of the repositioning of culture in development thinking and practice (Chapter 1, this volume; Watts, this volume) and sustained indigenous protest over the past two decades, indigenous engagements with development have recently been transformed, although not completely to many Indians' satisfaction. Indigenous development targets Indians as project beneficiaries, and often includes them in the design or implementation of projects. Celebratory accounts of indigenous development view it as a way in which "participatory planning can contribute to decentralized decision-making; stimulate the participation of the grassroots in local planning, help rural communities formulate development strategies, and increase the sustainability of investment" (Uquillas 2002: 14). However although indigenous peoples' specific cultural and institutional contexts have been awarded greater acknowledgement in development thinking and practice, indigenous empowerment and secure access to livelihoods remain difficult to establish and guarantee. As we see below, indigenous development measures do not necessarily permit Latin American Indians to set the terms of interventions. One reason for this is that current development policy confines indigenous groups to restricting places, activities, and relationships with non-indigenous groups. Our chapter's argument is that, despite efforts to rethink the relationship between indigenous people and development, Indian culture is incorporated into development thinking and practice within particular imaginative geographies of policy.[4] These developmental imaginative geographies fix indigenous development to limited scales, spaces, and social groups reflecting a spatially fixed vision of culture as localized, ethnically homogeneous, and founded on gendered divisions of labor. The concept of spatial fix originates in Marxist geographers' critiques of capitalist political economy, where capital seeks a spatial fix for its economic crises by disinvesting in one region and investing heavily in another place in dynamic topographies of redevelopment. One response to such processes is an increase in place marketing, as the uniqueness of certain places – expressed in cultural economic terms – are promoted to attract funding away from other areas. In the context of indigenous development, the commercialization of artisan goods or indigenous services (such as ethno-tourism and eco-tourism) attempts to attract foreign tourist capital and in turn diversify vulnerable rural livelihoods. In order to maintain profit levels, capital cyclically creates new built environments for expansion, while leaving other areas behind, thereby producing a constant process of creative destruction and uneven development (Smith 1984). Despite their distinct intellectual origins, we combine the concepts of imaginative geography and spatial fix in creative tension to provide a materialist critique of cultural relations, and a post-structuralist take on political economy. Together the concepts provide an insight into Andean cultural economies where development thinking and practice are making productive use of the category of indigenous culture. Drawing on examples of indigenous development from Andean countries, our chapter explores the conditions under which development can bring greater benefits to indigenous peoples as well as the limits on development's transformation.[5]

The chapter is structured as follows. First, we outline the diverse positions of Andean indigenous people vis-à-vis development, contrasting rare successful examples of indigenous controlled socioeconomic improvement with the majority experience of poverty and marginalization. In the second section, the chapter examines how development thinking and practice has transformed the treatment of indigenous culture in recent decades, particularly in its celebration of Indian cultural specificity. A more detailed analysis of projects in the countries of Ecuador and Bolivia in the third section permits a critical evaluation of the implications of such development thinking and practice for Indian people. Moving back to Latin American examples, the following section explores the socio-spatial fix by which developmental imaginative geographies shape the kinds of indigenous development available to transform Indian well-being.

Indians in development: diverse positions and opportunities

Due to the uneven landscapes of development across Latin America and differentiated experiences under colonialism and republican rule, indigenous people are positioned in political economies and projects in highly diverse ways. Small-scale Amazon societies' recent engagement with global oil companies brings them into very different development interactions to those of highland Indian peasants whose labor has been exploited by resource extraction (such as mining) as well as commercial agriculture over five centuries. Despite these disparities, highland and lowland Indian groups come together to propose improvements to government policy as exemplified by six Ecuadorian indigenous organizations' coordinated contestation of the neo-liberal Agrarian Development Law in the early 1990s. In spaces such as these, indigenous activism has worked across ethnic difference as well as frequently across national boundaries, making them interlocutors with donor, NGO, and government agencies where issues of culture and development are considered. It is possible to identify patterns of interaction between development and indigenous cultures, in order to distinguish between conditions under which Indian groups perceive benefits from and control over development, and situations of development lack. In other words, although the majority experience of indigenous groups is one of poverty and cultural marginalization, there are situations within which improvements in living standards are compatible with retention of an ethnic cultural identity.

Where – unusually – Indian groups have retained management of local resources and economies, they have been able to make social and economic gains while maintaining dynamic and resilient cultural identities. Independent craft production with Indian control over marketing and proceeds of sales is based upon secure tenure to sufficient land through the twentieth century and investment in community institutions (in terms of time, commitment, and money). The combination of independent craft production, non-capitalist social relations (including reciprocal labor and goods exchange), and high levels of ritual activity (supported by gains from capitalist economies) has beneficial impacts on cultural and development indicators (Stephen 1991). Whether among the Otavalo weavers of Northern Ecuador, or the Nahuatl bark painters in Guerrero state, Mexico, high levels of household and community involvement in market economies as producers and/or merchants is balanced by active, sustained contributions to cooperative activities and non-market exchanges. Among Zapotec weavers in Mexico, for example, three-fourths of households contributed to, and drew upon, community-based interest-free loans of labor, cash, and goods, and actively participated in annual ceremonial cycles through which social ties were strengthened (Stephen 1991: 116). In the Bolivian Andes, Indian households treat community activities – specifically ritual events – as crucial to reproduction, spending around 10 percent of their income on these items while simultaneously engaging in migrant labor and product sales (Platt 1982: 45). Moreover, reasonable access to markets is a crucial precondition for indigenous productive enterprises (North and Cameron 2000). Where sufficient levels of infrastructure, information, and credit exist, Indian producers have been able to overcome problems of market access and control. In this context, Bolivian indigenous women for example have created international trade links for marketing their medicinal products (Sikkink 2001). The combination of independent forms of production and access to credit via indigenous kinship networks also underpins indigenous garment workshops producing fake designer jeans in Cochabamba, Bolivia. Here, the resilience of Quechua cultural identities occurs via the transfer of skills from rural to urban places, and the social reproduction of key events in the rural communities' cultural calendar (Laurie 2005). A comparison across 28 self-generated Latin American indigenous projects identified ten preconditions for success, some of which have been mentioned above. The preconditions include basic human rights, food security, secure land and resource rights, indigenous participation in planning and implementation, intercultural education, strengthening Indian civil organizations, diversification of production, appropriate financial and technical assistance, and state support for indigenous self-development (Griffiths 2000: 16–17). When these factors come together, indigenous opportunities for self-managed and culturally enhancing livelihoods are greater; the rarity of this combination of factors too often results in material deprivation.

Although highly diverse in language, ethnic identity, livelihoods, and political perspectives, the most likely development situation for an indigenous person to find her/himself in is one of relative marginalization and exclusion from development benefits. In the world's most unequal region, Latin America's Indians are found among the poorest of the poor. In the Central Andean countries of Ecuador and Bolivia, Indians are on average among the poorest groups, a situation reflected in low levels of monetary income, poor health and education indicators, and low life expectancy. Around 64 percent of Bolivia's Indians are under the poverty line, compared with around 50 percent of the total population (Davis 2002). In Ecuadorian parishes with an indigenous majority, around 85 percent are poor, making these places 32 percent poorer than the national parish average and 14 percent poorer than the average rural parish (Uquillas 2002: 5). In Mexican municipalities with large numbers of indigenous people, four-fifths of residents live in poverty (Pirttijarvi 1999). The reasons for Indian marginalization are numerous. The general lack of access to education is a common pattern for many indigenous individuals, due to the underfunding of rural and urban Indian areas and widespread racism in curricula, policy, and teaching staff. In Bolivia, indigenous individuals receive three years less schooling on average than non-Indian people, while in Peru, indigenous adults have one-fifth fewer years schooling than their non-indigenous counterparts (Pirttijarvi 1999: 9).

High levels of poverty and marginalization reflect Indians' frequent subordination to unevenly developed political economies and their political exclusion from national and agency decision-making processes. An agricultural census undertaken in Ecuador by SwissAid together with Indian confederations found that three-quarters of the country's basic foodstuffs came from small, largely indigenous, producers yet they controlled only seven percent of irrigation water (Authors' interview 2000). Global commodity chains involving Indian producers are often long and weighted towards end consumers. Ecuador's Otavalo weavers thus receive only 1/400th of their products' final sale price when exports go via middlemen to the United States (Meisch 2002). Guatemalan Indian women weavers from San Pedro Sacatepequez were no better off in the 1990s than they had been in the 1970s when they first started producing cloths for sale (Ehlers 2000). Andean rural producers, many of them indigenous, provide indirect subsidies to urban consumers as they use more of their own labor and less capital than capitalist farmers, reinforcing the undervaluation of Indian labor and generating only low prices for their food products (Lehmann 1982).

Indian groups have often been sidelined by state and international development programs as their "culture" – often defined in closed, static, and one-dimensional ways – has been blamed for their inability or unwillingness to engage in the activities associated with modern progress. In Latin American republics, Euro-American sociocultural features have been consistently valued above indigenous elements, a perspective long underpinned by racist explanations for Indian "backwardness" (Graham 1990).[6] Development agendas have thus been profoundly shaped by the regional racial formations within which they are debated and applied (White 2002; also Chapter 1, this volume on racial formations). Elite non-indigenous groups often expropriated indigenous communities' land while tying Indian labor to agricultural and workshop production by harsh debt peonage. During the twentieth century, Indian cultures and economies were treated as "problems" that required a combination of cultural change, assimilation into nationalist schooling and identities, and greater involvement in capitalist economies as consumers and laborers (Larson and Harris 1995). As many Indians were located in rural economies, mid-twentieth century development projects treated them as peasants – rural producers whose productivity and agricultural techniques were to be improved along Western lines (Escobar 1995).[7] Due to poor rural educational provision and limited political representation, indigenous groups had only restricted opportunities to take control of development projects and local political economies. For example, agrarian development funds and social development funds are difficult for indigenous people to access (Authors' interview, 1999). Faced with racism, assimilating national cultures, and a lack of rural opportunities, many Indians abandoned indigenous identities and rural areas. During periods of rapid rural-urban migration during the mid-twentieth century, Indian families and individuals took advantage of more open urban economies and societies to rework social and cultural identities away from the stigmatized Indian category (Rowe and Schelling 1991; Larson and Harris 1995). It was widely assumed in Latin America, as in other world regions, that indigenous cultures would disappear as development progressed (Hettne 1996: 15).

During the 1980s, however, a number of factors came together, prompting a rethink of indigenous culture and the role of Indian groups in national and economic life, in the context of shifting development paradigms. While some of these factors have their counterparts elsewhere in the world, some were specific to Latin America (see Chapter 1, this volume).[8] The first of these factors was Latin America's 1980s reversal of development known as the "Lost Decade" when economic downturn and rising international debt levels caused increasing levels of poverty, unemployment, and worsening health and educational measures for most social groups. Disillusioned with modernization paradigms, Latin America went through a development crisis in the wake of which highly diverse options were pursued. While certain actors vaunted neo-liberal asset privatization, state budget cutbacks and globally oriented production (Dezalay and Garth 2002), others turned to romantic views of Indians, while others advocated "local authentic solutions," religious revivals or postmodern skepticism (Larraín 2000).

A second factor was Indian political mobilization across a number of countries to demand more secure land titles, decision-making power, and an end to discrimination (Brysk 2000). Although some Indian federations had been established in earlier decades – Ecuador's Shuar Federation was founded in 1964, for example – the 1980s saw the emergence of more numerous organized groups, an educated leadership, and growing coordination at national and international levels (Brysk 2000). In the wake of peaceful civil protest by indigenous organizations and other movements, nation-states established constitutional and legal rights to bilingual education, means to register land collectively and to cultural recognition (Assies, van der Haar, and Hoekema 2001; Sieder 2002). Most Latin American countries enacted varying degrees of economic restructuring alongside multicultural reforms, which gave indigenous populations contradictory positions within which to rework or resist economic, social, and political reforms (Hale 2002). As a consequence of these simultaneous reworkings of Latin American cultural economies, many indigenous people became subjects that engage the cultural project of neo-liberalism: "neo-liberalism did not undermine indigenous cultural difference . . . but instead engaged Indian difference to symbolize a different kind of subject – the Indian citizen – a subject who acts as a global rational being with rights and needs" (Valdivia 2005: 290).

The third factor was the increasingly formalized set of international measures to protect the world's indigenous groups and their human rights (Tennant 1994; Brysk 2000). The United Nations and other multilateral bodies began to redress indigenous populations' unequal access to rights and security. Ratification of International Labor Organization (ILO) convention #169 by a number of Latin American states established a baseline for the inclusion of indigenous development agendas, as it established Indian rights to collective territory, decision-making powers over development, and respect for cultures (Brysk 2000).

As development projects that ignored the specific problems and concerns of indigenous people often failed, the events outlined above provided a new impetus to listen to indigenous people when they offered alternative development proposals. Non-governmental organizations (NGOs) and human rights networks who had been working alongside indigenous people for more culturally appropriate forms of development found themselves consulted and recruited into national and international initiatives (Andolina, Laurie, and Radcliffe forthcoming). In Ecuador and Bolivia, indigenous organizations protested against harsh restructuring policies and called for a greater say in development planning and implementation (Pacari 1995). In Colombia's Pacific Coast region for example, logging, hydroelectric and mining activities damaged indigenous and Afro-Colombian livelihoods and the environment. Mobilizing a range of support organizations, diverse ethnic groups put together a series of proposals for indigenous and Afro-Colombian participation in decision-making and socially and environmentally sustainable development (Atkins and Rey-Maquieira 1996). In the northern Ecuadorian province of Esmeraldas, the custodial rights and management practices of Afro-Ecuadorian groups to mangrove swamps have been recognized in a new ecological reserve (Ocampo Thomason in press).

During the late 1980s and particularly from the 1990s, bilateral and multilateral development agencies began to overhaul their policies systematically to take indigenous peoples into account. National governments wrote policy papers on indigenous development in order to direct overseas aid and assistance practice (countries included Denmark, Norway, United Kingdom, the Netherlands, Finland). In line with other European approaches, the Dutch policy is informed by the wish to hold national governments responsible for indigenous peoples' treatment and to maintain Indian cultural identity (van Schaik 1994). Responding to Latin American Indian confederations and indigenous government representatives, an international "Indigenous Fund" was established in 1992, based in the Bolivian capital La Paz, to make grants for indigenous development from a large endowment. Similarly, multilateral development agencies such as the World Bank, and Inter-American Development Bank established specialist groups and formalized policy on indigenous issues. The World Bank's operational framework on indigenous people, known as OD 4.20, shifted policy from a reactive to a proactive mode whereby projects affecting indigenous groups would work for informed indigenous consent and targeted development projects. Between 1992 and 1999, one sixth of the Bank's Latin American portfolio was in indigenous-related projects in environmental, social, rural, and human development (Davis 2002: 234). Despite draft revisions of the Bank policy in 2004, after a decade's application of OD 4.20, indigenous and human rights groups remained critical of how multilateral policy gave Indians no right to self-identification nor to prior and informed consent to projects (Bank Information Center 2004).

Development and indigenous culture: changing paradigms

From the late 1980s, and with increasing speed, development began to take indigenous people into account to an unprecedented extent, seeking to bring economic development and cultural goals into harmony. In the words of the region's major international development agency, "strengthening cultural identity and sustainable development are mutually reinforcing" (Deruyttere 1997: 9). In Latin American culture and development thinking, the tendency has been to equate *culture* with *indigenous* culture, with both neo-liberal and Marxist economists viewing "traditional" cultures as positive instances of cooperation and as a means of resisting globalization (Hojman 1999: 175; Carranza 2002; compare Chapter 1, this volume). In the (often ambivalent) contrast between a Western-oriented modern national culture – including appropriations of Indian folklore (Bigenho, this volume) – and "backward" indigenous cultures, the search for cultural difference and authenticity leads back to indigenous forms (Rowe and Schelling 1991).[9] In line with wider debates about globalization's threat to cultural diversity (see Chapter 1, this volume), the protection of "indigenous languages, cultures, values [and] use of indigenous knowledge" (Davis 1999) underpinned World Bank and environmentalists' responses to dam building and similar development. By contrast, many indigenous actors view themselves as "shuttling between discourses" (Valdivia 2005) as well as cultural and development identities, seeking to make connections between cultural expressions and neo-liberal development and constructing multiethnic coalitions, rather than treating Indian cultures as merely vulnerable pockets of difference.

Development's profound shift in the treatment of indigenous cultures has coincided with the dominance of the neo-liberal development paradigm. This is not the place to go into a detailed examination (see Watts, this volume), but suffice it to say that neo-liberal development thinking shapes the context for indigenous development paradigms and practices, albeit in highly diverse, uneven, and often contested ways (Larner 2003).[10] Drawing on a wide repertoire of concepts and practices, neo-liberal development extends its core concerns with dynamic market economies, reworked governance, and deregulation into indigenous development. It views indigenous populations as exemplary representatives of the power of social capital, the dynamism of small-scale entrepreneurialism, and non-state forms of authority and leadership (Radcliffe, Laurie, and Andolina forthcoming; also Watson, this volume). Acknowledging the persistence of indigenous poverty, neo-liberal approaches have begun to target programs at Indian groups (and at other poor groups) and to recast development's racial formations around indigenous difference. Celebrating examples of indigenous "can-do" attitude, the current mainstream view argues that indigenous development

> is most likely to occur when indigenous peoples have access to basic resources for their social reproduction, including food security and basic health conditions; have achieved a high degree of social organization and political mobilization; have been able to preserve their cultural identity; have built strong linkages with outside organizations; and have productive patterns that allow for subsistence and earning cash income.
>
> (Partridge and Uquillas 1996: paragraph 20)

Social capital, that is the mobilizing networks and relations of trust
that underpin a sense of community, has been identified as one of
Latin American indigenous groups' key resources (Radcliffe, Laurie,
and Andolina forthcoming; compare Loomis 2000). According to one
multilateral assessment, "social exclusion, economic deprivation and
political marginalization are sometimes perceived as the predominant
characteristics of Ecuador's indigenous people. But as they often
remind others, indigenous people also have strong positive attributes,
particularly their high level of social capital" (van Nieuwkoop
and Uquillas 2000: 18). Neo-liberal development presumes that
social capital strengthening will expand other forms of capital while
retaining indigenous groups' cultural specificity (Davis 2002: 232–3).

Neo-liberal development thinkers argue that Indians' exclusion from
development has negative economic impacts, and that consequently
their inclusion in development will have beneficial impacts not
only on Indian incomes but also on economically unstable national
economies. According to a major World Bank report on indigenous
development, Indian exclusion from the market economy represents
"a massive waste of human and non-human resources; their inclusion
would lead to a productivity boost, strengthen purchasing power and
promote growth" (Partridge and Uquillas 1996, paragraph 17).
In the context of neo-liberalism, indigenous development is often
framed as one component of a vigorous market economy, freed
from the shackles of the state and embedded within strong social
norms. Accordingly, cultural diversity becomes an engine for
economic growth as it is intrinsically connected to the market and
has the additional advantage of operating within culturally specific
institutional arenas outside the state (Radcliffe and Laurie 2006).

Within this policy approach, indigenous groups' contribution to
development is to come from their economic activity. Neo-liberal
development emphasizes the need to make Indian goods and
services available to markets and encouraging Indian groups'
business acumen. As part of boosting of non-traditional exports,
products such as organic cocoa, *quinoa* (an Andean grain),
distinctive clothing, and wood products are encouraged by
multilateral development projects. In 2004, for example, the
Inter-American Development Bank (IADB) granted Guatemalan
Indian producers of organic fair trade coffee US$461,000 to
boost exports.[11] The IADB views the small-scale coffee farmers'
association Manos Campesinas as a prime example of social
entrepreneurialism, as the community-based enterprise combines

"sound business practices with a strong social commitment" (IADB 2004). Likewise, Latin American governments together with non-governmental organizations and international development agencies have initiated programs of ethno-tourism, eco-tourism, and "traditional" music exports in order to boost the earnings from/for indigenous groups. Although less market-oriented than neo-liberal ones, NGO-led projects also encourage indigenous populations to develop culturally distinctive activities, although these approaches often additionally have an empowerment agenda which engages more in challenging racist exclusion. However, the latter approaches tend to receive less funding and prominence in debates than neo-liberal programs. Indian groups are encouraged to develop ethno-agriculture, organic agriculture, indigenous crafts, inter-cultural education, and non-mainstream media and communication (Kleymeyer 1995; Healy 2000; compare Kaplan 1995). Given the exclusion of many Latin American indigenous populations from core citizenship rights, there is much mileage in addressing the highly uneven coverage of health, education, banking, and technical assistance services in majority indigenous areas, issues addressed by indigenous political and human rights organizations (Stavenhagen 1996; Molyneux and Lazar 2002).

Pro-indigenous development in Ecuador and Bolivia: the end of marginalization?

[Andean Indians] really maintain their culture, their own identity. There's really a lot of vigor in their culture, their technologies, their wisdom.

(Authors' interview with major international
NGO, Lima 2000)

Indigenous development projects in Ecuador and Bolivia illustrate the parameters of development thinking about indigenous culture, while also permitting an analysis of the implications of this version of culture and development for beneficiary populations. In Bolivia and Ecuador, indigenous development has galvanized the support of a wide range of development actors from local indigenous village associations through to multilateral agencies. However any celebration must be tempered by the recognition that the multiple actors involved in "culturally appropriate" projects have widely varying capabilities to shape the direction and extent of change, and the multiple geographies it engages. With large numbers and percentages of indigenous populations, Ecuador and Bolivia grapple with the difficulty of profoundly reorienting a racialized development trajectory in more inclusionary ways and attempting to control an increasingly globalized economy around the diverse interests and cultures of their multiple ethnic groups. Around 40 percent of Ecuador's population speak an Indian language, self-identify as indigenous and/or live their lives by indigenous social organization, in one of nine major groups. In Bolivia, Indians comprise 59 percent of the population, in 36 linguistic groups, although Aymara and Quechua-speaking people comprise 25 and 33 percent of the highland population respectively. Andean indigenous development engages highly diverse actors, paradigms, and geographies although in the two countries examined here, the main actors are the nation-state (often in conjunction with multilateral development agencies), non-governmental organizations, and indigenous organizations and communities.

Across the range of Ecuadorian and Bolivian indigenous development projects, two major approaches can be identified. On the one hand, culturally distinctive products and services are brought into development agencies' promotion of non-traditional exports and entrepreneurs, while on the other hand "traditional authorities" and institutions are rediscovered as the basis for reworked forms of governance (Radcliffe and Laurie 2006; also Watson, this volume). Indigenous cultural difference is recognized in Andean countries' development yet it is visible very largely in the terms constructed by the "Indian citizen" slot. Indigenous cultural economies are perceived in policy as culturally embedded yet are interpreted as incipiently capitalist:

> So in my view the indigenous people can actually enter the market
> and be successful, although it doesn't mean they follow the Protestant
> model of capitalist accumulation or into individualized businesses;
> they can follow a more community-oriented business.
>
>   (Authors' interview, La Paz, 2000)

Responding to international negotiations with indigenous peoples and non-governmental organizations, the Bolivian government passed indigenous rights legislation and made "development with identity" the basis of Indigenous Ministry policy. Given that indigenous development went national at the same time as governments established profound neo-liberal economic and social restructuring, it is no surprise that the predominant model adopted was a market-oriented and institutional one. Preparatory documents detail how indigenous development needs to explore the "economic advantages of the live indigenous culture, whether native or vernacular, as it applies to urban and rural cultural goods and services . . . and religious and folkloric practices. [. . .] Other expressions of live indigenous culture are crafts, weavings, pottery, wood carvings, leather work, and Amazonian products" (World Bank personal communication). Indigenous development was to be the cornerstone too of participation and decentralization processes (that is, politico-administrative reform) (Calla 2001). In contrast to previous attempts to make Indians into Western class actors as rural farmers, agrarian development funds now allocate 10 percent of budget to projects of cultural recuperation (Andolina, Laurie, and Radcliffe forthcoming). Bolivia's new Learning and Innovation indigenous development Project aims to build on indigenous knowledge to generate projects in ethno-tourism, ethno-biology, and agriculture, all of which are designed to be marketable and competitive (World Bank personal communication, 1999). Following the agendas outlined above, many large donor inspired projects focus on promoting indigenous human capital and expertise through various professionalization programs (Laurie, Andolina, and Radcliffe 2005).

A similar range of indigenous development initiatives exist in Ecuador, where the nation-state has coordinated with Indian confederations and multilateral agencies to put together attempts at more culturally appropriate development specifically involving or targeting Indian populations. A number of examples illustrate the range of experiments that emerged in this context. Among the Amazonian Napo group of Quechua Indians, a Canadian-funded local NGO attempted to promote indigenous culture via a museum, ethno-tourism and ceramic crafts. Working from the NGO director's view that "viable cultural practices are those that integrate successfully with the market," the project clashed with indigenous concerns about the ownership and benefits from this commodification (Wilson 2003: 168). The revaluation and promotion of material cultures was further endorsed by the Inter-American Development Bank, which granted US$1 million to Ecuador's indigenous ethnicities to promote their entrepreneurial capacities.

Phase One of the Development Project of Indigenous Peoples and Afro-Ecuadorians of Ecuador (PRODEPINE I)[12] funded the recuperation of indigenous culture, which is equated with festivals, rituals, and archaeological sites, during the mid-1990s. Funded by the World Bank, Ecuadorian government and IFAD (International Fund for Agricultural Development) to the tune of US$50 million, PRODEPINE I funded festivals, video work, and crafts to "rescue and strengthen a rich cultural patrimony." Over the project's lifetime, PRODEPINE I supported cultural identity and strengthened indigenous organizations, as well as benefiting around 1,300,000 indigenous women and men (IFAD 2004). One of PRODEPINE I's central aims was to recuperate and spread knowledge about indigenous culture within Indian groups, thereby strengthening ethnic identity and cultural forms. In workshops on culture, participants discussed and recorded (in written and video formats) different groups' languages, clothing, festivals, forms of reciprocity and solidarity, beliefs, burial rituals, social sanctions, music, and dance. With the agreed strategy of support for the handicraft and tourism sectors, the project ran concourses to establish quality standards and held intercultural fairs and events, where services and products "can be displayed and promoted" (van Nieuwkoop and Uquillas 2000: 21–2). The project's cultural component supported the production of four documentaries, 28 festivals of music, and three workshops on cultural inheritance, yet was evaluated as unsatisfactory as the "aim is to insert culture well into income generating activities and productive investments" (Uquillas 2002: 15), and income did not match expectations. One project element was the creation of trademarks, labels, and certificates to highlight the quality and distinctive cultural origin of goods and services but was little used (van Nieuwkoop and Uquillas 2000: 22).

In the second phase of PRODEPINE, funded to US$40 million and approved in June 2004, successful project components are programmed to continue alongside new elements founding the project more securely upon the notion of cultural difference.[13] Community investments, land-titling, scholarships, and institutional strengthening, perceived as strong parts of the first phase, are to be extended while new work in natural resource management and cultural development defined in terms of community values and cultural assets are to be added (ITDG 2005). The number of targeted beneficiaries is set to increase to 1.5 million indigenous and Afro-Ecuadorian people (World Bank 2004 press release), while more attention will be awarded to ethnic and cultural variations, with separate development plans for Indians and Afro-Ecuadorians respectively (Japan PHRD 2003). Central to PRODEPINE II is the training and technical assistance to communities engaged in handicraft production and marketing, community-based tourism, and production and marketing of products based on ethno-biological knowledge and resources (Japan PHRD 2003). Privileging culturally distinctive products and services, the project's second phase establishes indigenous groups more securely within a globalizing and differentiated consumer-led type of capitalist development.

Looking across the various examples of indigenous development – whether devised within neo-liberal multilateral development agencies or emerging from historically embedded patterns of indigenous autonomy and livelihood – the factor of indigenous security appears to be a crucial element. Otavalo traders in textiles have a marketable identity and social capital, yet income depends on secure access to a market stall and stable supply marketing chains (Kyle 1999: 436). The fields of biotechnology, bio-prospecting, intellectual property rights, and exploitation of medicinal plants offer insecure ownership (Authors' interviews New York and Quito, 2000). Despite many years of mobilization to clarify collective land rights and individual title, indigenous development projects can be slow and inefficient with regard to this means of security. PRODEPINE I's component of land registration was slow to get off the ground and relied heavily on a national NGO that had many years experience of working with Indian land claims, and project resources available to the NGO for this work are limited (Griffiths 2000: 33). Only after a decade of transnational social movement and NGO pressure did the World Bank revise its indigenous policy to recognize individual and collective land rights.

Indigenous groups are currently being encouraged to undertake certain types of activities which often engage them in sustained interaction with members of multiple and diverse ethnic, national, and racial groups. Given the histories of many indigenous groups, especially peasant populations in the Andean highlands, these interactions are not unprecedented. Indigenous women traders who rely upon tourist sales have long known the importance of an indigenous appearance in enhancing their "authenticity" and likelihood of success in both service and artisan sectors (Cone 1995; Crain 1996). Yet in the context of globalizing economic relations and Indians' growing reliance upon the external sales of services and products, these relations take on a new urgency.[14] The maintenance of a highly controlled and costly form of bodily self-presentation has become increasingly important in a wide-range of post-Fordist employment (e.g. McDowell 1997). Moreover, Indians' self-presentation risks restricting them to a devalued identity, as the racism that underpins the boundaries between Indian and other racialized identities often remains in place.

Development needs to build upon Indians' existing multiethnic and cross-cultural relations to challenge expectations of colorful ethnics and widen knowledge of the realities about Indian livelihoods and concerns. Quechua women in the Bolivian organization ASUR trade their weavings across the world, communicating with many diverse cultures yet rely upon working over four months to raise US$100 per item (Healy 2000). ASUR's creative work to "use cross-cultural education to adapt the market to the [indigenous] artisans" (Healy 2000) illustrates how empowering intercultural practices and respect can develop from multiethnic encounters (see also Bigenho, this volume; Nederveen Pieterse, this volume).

Indigenous development on the ground: socio-spatial fixes?

Using Latin American examples, we now explore how indigenous societies and spaces are represented in policy and then how development practices rework Indian development opportunities. Responding to economic and cultural concerns, indigenous development projects attempt to bring Indians more successfully into niche global markets and to establish the basis for cultural identity and strengthening of indigenous institutions and social relations. Indigenous development policy thus plays a role in reproducing the geographical, economic, social, and cultural relations through which indigenous groups are reproduced inside capitalist political cultural economies. Our question here is thus, to what extent does indigenous development practice in effect provide a "spatial fix" on indigenous development and, if so, in which geographies? What are the implications of these for indigenous well-being? In the following examples, we explore the sub-national regional imaginative geographies, the gendered fixes of indigenous development, and questions about urban indigenous groups.

The first risk in indigenous development is attempting to provide one solution for a highly diverse population, whose engagements in economic and cultural relations are often highly differentiated and even mutually incomprehensible. As the example of PRODEPINE I illustrates, a blanket policy can be founded upon imaginative geographies that mark out one homogeneous group of indigenous people as a norm. PRODEPINE I has garnered respectful assessments as much from its multilateral donors as from longstanding critics of targeted neo-liberal programs. However, commentators consistently point up the ways in which it works on the ground, shaping and reshaping the spaces of development and the relationships of diverse ethnic groups to the project. The project proclaims its commitment to "development with identity" and "interculturality" yet paradoxically uses the *same* procedures for *diverse* ethnic and cultural groups (IFAD 2004). Although founded upon the notion of (ethnic) difference, the project in practice does not take cultural, geographical, and political-administrative variability into account. Specifically, there is a widespread perception among project beneficiaries and commentators that the project was primarily designed for the Sierra region of Ecuador (that is, the Andean highland region). Indeed, the Sierra received 63 percent of total funding. Moreover, project elements were implemented more quickly in the Sierra than in the lowlands or coast (Griffiths 2000: 28). Thus although the project was designed to supercede historical associations of Indians with poverty and the Andean mountains, the imaginative geography of policy in effect has refixed the Andean Sierra as the place where indigenous development occurs (Radcliffe, Laurie, and Andolina forthcoming). Reflecting Andean Indian experiences of land insecurity and peasant economies, as well as regional forms of indigenous organization, this imaginative geography of indigenous development policy uses a specific interpretation of socio-spatial relations in other, very different contexts. As a consequence of the Andean policy template, Amazonian indigenous groups and Afro-Ecuadorian populations argue that the project does not address their core concerns nor provide them with the expected benefits. The Shuar group in Ecuador's Amazon found that public participatory workshops on development priorities clashed with culturally established processes of lengthy kin-based discussions (Larreamendy 2003: 185–6). Pressure groups and multilateral agencies have begun to call for specific multiple Amazon and Afro-Ecuadorian PRODEPINEs in order to address these shortcomings (IFAD 2004).

Another dimension of spatial fixing is the way in which indigenous women are treated in indigenous development projects. Indian women's development concerns are wide-ranging and challenging to any development agency. Compared with indigenous men and non-indigenous women, the majority of Indian women find it difficult to access education, training, secure employment, and adequate healthcare and bodily well-being.[15] Depending on specific local gender divisions of labor, rural-based Indian women can find themselves undertaking more agricultural work when men out-migrate, or women are the first to migrate into urban areas (often as domestic servants) as in many Ecuadorian highland areas. In Chile too, Mapuche women are more likely to migrate to cities than Mapuche men, although they then face "hard poverty" due to poor qualifications, racism, and limited job opportunities beyond domestic service (González 2003). An underlying expectation of much development thinking is that women form the basis of a culturally and spatially distinctive rural indigenous community, an expectation that profoundly shapes policy and practice (Radcliffe, Laurie, and Andolina 2004). Many policy-makers and Indian leaders represent women as actors who maintain an ethnic group's cultural distinctiveness by rearing children, speaking indigenous languages, and wearing ethnic clothing.[16] Asked why their organization focuses on indigenous peoples, one international NGO staff member replied:

> The most marginalized group, the most dispossessed are the Indians. In this large group, the second thing is . . . Look – it's a group that can be considered as a reserve really, the possessor of ancestral cultures and in relation to this, particularly the women.
> (Authors' interview 2000)

Reporting on the nature of Ecuadorian indigenous social capital, a World Bank policy document argued that women are more likely to stay in communities with high levels of social capital (World Bank, personal communication 1999), thereby associating women's work in sociocultural reproduction with localized settings.

Development projects then devise mechanisms for indigenous women that reinforce or build upon these preconceptions, if indeed they take gender into account at all (see Radcliffe, Laurie, and Andolina 2004)! In PRODEPINE I, women were awarded micro-credit facilities that built up locally oriented productive activities in women-only groups based in villages (Tene, Tobar, and Bolaños 2004).[17] Under the project indigenous women were granted capital of around US$500 per village for savings schemes (*cajas solidarias*), compared with the average of US$15,000 for a "general" village project (Authors' interview 2000). By granting women such small amounts of capital, the project reinforced a localized set of gendered activities while working against women's diverse itineraries of migration, and travel for trade and exchange (Authors' interview 2000). While no doubt raising women's confidence and expanding domestic small animal production, the reinforcement of a women-only "add on" fails to address women's lack of power to decide on broader dimensions of indigenous development, gain access to land, and further careers. Diverse Ecuadorian Indian women highlight these latter development issues as central to their demands, yet they are rarely incorporated into mainstream indigenous projects. Evaluating PRODEPINE I, IFAD "could not verify the existence of a gender strategy that promotes women's decision-making and opportunities for training, business management, lending for land purchase and legalization, scholarships and other project interventions" (IFAD 2004).

A final aspect of indigenous development's spatial fix concerns the silences in policy about urban-dwelling indigenous people and national cultural economic relations. Under the existing indigenous development paradigms in operation in Latin America, Indians are represented as rural, often peasant, producers and the existence of large numbers of indigenous people in the region's cities and towns is downplayed or ignored completely. As one Bolivian indigenous leader commented, the Aymara people "includes urbanite professionals, not only *campesinos* [rural dwellers]." Although policy-makers *do* picture Indian merchants in urban economies, they are seen as exceptionally entrepreneurial and hence not representative of the majority of Indian populations. Indigenous organizations have criticized policy for this rural bias, for example, by drawing attention to the lack of an urban dimension to the World Bank's revised policy on indigenous development (Bank Information Center 2004). Indigenous leaders astutely point out the inconsistencies of multilateral development policy for failing to address Latin American societies as multicultural and racialized. In a meeting between highland indigenous leaders and counterparts from Guatemala, Bolivia's bilingual intercultural education program was critiqued for being applied "only in the countryside, not in the city," thereby reinforcing the association of culturally distinctive Indian subjects with rural areas and non-Spanish fluency (Fieldnotes 1999). By separating out distinctive Indian rural areas, many indigenous projects reinforce an understanding of indigenous livelihoods as distant from wider political economic processes and actors. The inability of small-scale community projects to intervene in neo-liberal development policy at the national and international levels is highlighted by Indian organizations, independent commentators, and project personnel (Authors' interviews 1999–2004). Although donors view PRODEPINE II as offering high strategic rewards, the project's annual funding of US$40 million for 2004 and 2005 will be dwarfed by neo-liberal competitiveness and structural adjustment loans which will total some US$100 million *annually* in 2004–7.

Conclusions

The cultural economies of Andean indigenous peoples exist in the interstices of uneven developments and exclusionary racial formations within which the combination of security, access to resources and markets, and recognition as full citizens is rarely achieved. The resultant severe consequences for the development opportunities of the majority of the region's Indians have recently been taken on board by policy-makers who highlight the potential offered by cultural difference as the basis for a distinctive development strategy. Yet the imaginative geographies of policy concerning Andean indigenous peoples – as local, rural, and culturally distinctive – tends to reinforce the exclusionary features of existing cultural economies. Although concerned about the potential loss of Indians authentic cultures under globalization, indigenous development policy tends to offer the global market as the arena through which cultural difference is to be reproduced. The spatial fix of capitalist political economies continues to tie Indian groups to ethnic homogeneity and an arena outside the mainstream, articulated around the model of the "Indian citizen." The Indian citizen embodies the goals of good governance (under the leadership of traditional authorities) and market-led development (ethnic crafts and tourism). Imaginative geographies of policy and the spatial fix thus combine to create a circularity of argument – a local, ethnically homogeneous, and rural Indian cultural economy will give rise to a (profitable and governable) local, ethnically homogeneous, and rural Indian cultural economy.

Under indigenous development projects, indigenous people are often highly motivated to succeed but continue to experience precarious working and living conditions as a result of low wages, variable working hours, and little long-term security. However valuable Indian services and products become in global markets, indigenous development is currently operating under the sign of restructuring where indigenous "freedom" to become business people occurs under forced adoption of the position of "entrepreneur" albeit without secure access to credit, institutions or cultural capital (see also Ellmeier 2003; Bigenho, this volume). Particularly in the context of mainstream neo-liberal thinking where culture is reduced to product and institution (see Chapter 10, this volume), the challenge is to maintain a flexible and dynamic concept of culture in development thinking and practice. Whereas early and localized experiments in Andean indigenous development built upon the creativity of popular culture (Kleymeyer 1995), the challenge has been to balance instrumentalist and creative aspects as indigenous development has been scaled up and reworked in neo-liberal spaces and institutions.

While recent policy represents a considerable advance on cultural economies that marginalized Indian labor, cultures, and societies, this comprises a limited interpretation of Indian culture and a misinterpretation of indigenous cultural economy. Separating economically beneficial behaviors from the sociocultural contexts that validate and underpin them works against the synergy of cultural identity and development opportunities. If the balance between self-exploitation and (economic and cultural) reward is forced into low rewards and exploitation by others then development activities have little long-term potential. As noted in relation to successful *self-generated* indigenous development, Indians' engagement with the social and economic benefits of new activities can work well if the latter fit with an existing life. Skills in production and marketing are not built up overnight, while the sustainability of production, trade, and innovative responses to development opportunities relies upon the maintenance – not the destruction – of culturally meaningful social networks and events. The challenge in future development programs revolve around the need to move away from indigenous societies as automatically comprising a culturally distinctive and geographically discrete group whose impulse is to produce goods and services for global consumers.

Acknowledgments

This chapter draws on research funded by the Economic and Social Research Council (Award no L214 25 2023).

Notes

1   Following Chapter 1, development is defined as the reworkings of relations of production and reproduction, and sociocultural meanings, in both planned interventions and the uneven patterning of political economy.

2   "Indigenous communities, peoples and nations are those which, having a
      historical continuity with pre-invasion and pre-colonial societies that developed
      in their territories, consider themselves distinct from other sectors of the society
      now prevailing in those territories or part of them. They form at present non-
      dominant sectors of society and are determined to preserve, develop and
      transmit to future generations their ancestral territories, and their ethnic identity,
      as the basis of their continued existence as peoples in accordance with their
      own cultural patterns, social institutions and legal systems" (Martínez Cobo
      1987, quoted in Assies, van der Haar, and Hoekema 2001: 4). As the region's
      census rarely include questions on ethnic-racial grouping and in the context
      of indigenous identity's stigmatization, total numbers are based on informed
      estimates. Indians make up between 8 and 10 percent of the region's
      population, with 85 percent found in Meso-America (Mexico and Central
      America) and the Central Andes (Ecuador, Peru, Bolivia, and Colombia). In
      this chapter, the terms indigenous people and Indians are used interchangeably,
      although many Indians prefer to be known by their language or cultural group,
      such as Quechua or Shuar. On the distinct difficulties of defining African
      indigenous peoples, see Barnard and Kenrick (2001).

3   To attribute indigenous populations with specific cultures is problematic,
      given their long-term interaction with – and mutual influence of – non-
      indigenous cultures. Nevertheless, as the definition of indigenous peoples in
      the previous note indicates, these groups often maintain distinctive relations
      with environments, social relationships, justice systems, and beliefs, more
      or less inflected by national and global patterns of meaning. Latin American
      indigenous cultures are highly diverse, varying with language group, patterns
      of (international) migration, and rapidly changing horizons of cultural reference,
      and structures of feeling (Stavenhagen 1996; Assies, van der Haar, and
      Hoekema 2001).

4   Edward Said's concept of imaginative geography refers to the ways in which
      Others' places are imagined as separate, distinct and distant. Said provided a
      post-structuralist critique of the imaginative geography of Orientalism (Said
      1978).

5   In this chapter we do not have space to discuss the diverse experiments in
      recuperating autochthonous crops and indigenous forms of agricultural
      techniques that occurred in the 1970s and 1980s in several Andean countries
      (e.g. Healy 1996).

6   The *indigenista* movements of Mexico and the Andean countries in the early
      twentieth century revalued and celebrated pre-Colombian Indian achievements
      (associated particularly with the Inca and Aztec Empires) and rewrote the
      narratives of nationalism but generally failed to challenge racist exclusion and
      poor Indian living conditions.

7   For these reasons, recent development policy has encouraged the re-emergence
      of the term "indigenous" to replace the class-based and state-oriented term
      "*campesino*/peasant".

8   In Latin America, culture – as opposed to indigenous culture – has been
      revalued in development for a number of reasons. Culture has been seen as the
      basis for collective action, a route out of the crisis in economistic thinking, a
      tool in fighting poverty and low self-esteem, the source of economically
      favorable attitudes, as a bulwark against globalization and to release the
      energies of the informal sector (De Soto 1989; Kliksberg 1999; Hojman 1999).

9 As a consequence, Afro-Latin Americans and other ethnic-racial groups have been relatively invisible.

10 On neo-liberal development in Latin America, Gwynne and Kay (2004) provide a good introduction. Evaluations of recent shifts in neo-liberal policy and projects include Perreault and Martin (2005), Laurie, Andolina, and Radcliffe (2005), and Bondi and Laurie (2005).

11 Coffee is of course a traditional Guatemalan export although the trade in organic and free trade products has a more recent history.

12 In Spanish, the project is titled *Proyecto de Desarrollo de los Pueblos Indígenas y Negros del Ecuador* (The Development Project of Indigenous and Black Peoples of Ecuador).

13 At the time of going to press, the second phase of PRODEPINE had been cancelled due to political events.

14 Elsewhere we discuss how project managers' and policy-makers' expectations of indigenous appearance shape their decisions about which groups in a locality become beneficiaries of an indigenous development project (Laurie, Andolina, and Radcliffe 2002).

15 It is extremely difficult to go beyond broad generalities about indigenous women, as – despite their incredible diversity and varied response to the conditions in which they find themselves – national governments and NGO projects rarely disaggregate national statistics by ethnicity and gender (see Vinding 1998; Radcliffe 2002).

16 By contrast, indigenous men are more likely to be represented as traditional *and* modern, bilingual, and able to move between Indian and non-Indian spaces and societies (see Radcliffe, Laurie, and Andolina 2004).

17 In Bolivia, the Banco Solidario also operated group credit schemes targeting women.

5 Laboring in the transnational culture mines: the work of Bolivian music in Japan

Michelle Bigenho

Since at least the 1970s Bolivian musicians have been touring Japan on short-term contracts, developing in that country one more niche market for an Other's music, much like the cases of jazz (Atkins 2001), blue grass (Mitsui 1993), salsa (Hosokawa 2002), hip hop (Condry 2001), or tango (Savigliano 1995). For Bolivian musicians, these flexible transnational contracts with Japanese companies offer a remuneration that is near impossible for musicians to match within their home country. In this chapter, my ethnographic lens focuses on the intersection of Japanese and Bolivian interests as I follow a Bolivian music tour in Japan and consider both the Japanese contracting company and the contracted Bolivian musicians. In a double sense, my analysis is about the work of culture: the laboring processes that break sweats and raise creative juices, as well as the symbolic work that occurs so that Bolivian music can be appreciated by Japanese audiences.

The work of these musicians can be located in the cultural economy of late capitalism, a context characterized by "the culturalization of economic life," and the privileging of cultural exchanges over economic and political ones (Ching 2001: 285). Cultural economy builds profits through the commodification of difference (Ching 2001: 288) and the cultivation of multiple differentiated niche markets (Jameson 1991; García Canclini 1992) that may provide limited economic benefits for those who supposedly embody and perform that difference. The primacy of a cultural economy comes with the late or post-Fordist capitalist shift from a production-oriented economy to a consumption-oriented one (Comaroff and Comaroff 2000: 295; Ellmeier 2003: 5)

While some approaches to cultural economy emphasize a media-generated circulation of cultural symbols, (Ching 2001; Iwabuchi 2002a), the examples I present emphasize cultural performances that are still at the margins of these media machines (on cultural economy, see Chapter 1, and James, this volume). I follow a Bolivian band that is somewhat well known in Bolivia, but that has remained outside of any glitzy commercialized national or transnational fame. When *Música de Maestros* (Music of the Masters) performs in Bolivia, they unabashedly follow a multicultural nationalist project, staging highland indigenous, lowland indigenous, and *mestizo* (mixed spanish-indigenous) musical forms. While staged indigeneity – that is, a presentation of indigenous identity which has been a part of the Bolivian music economy since at least the 1950s[1] – used to focus primarily on the use of three instruments associated with an imagined Indian world (pan pipes, notched flutes, and small plucked instruments or *charangos*), current Bolivian cultural politics demand specificity in the representation of different ethnic groups and an attention to perceived performance practices associated with particular groups and instruments. While the 1952 Bolivian Revolution brought the general inclusion of indigeneity into the national project, the specificity of today's musical representations is consistent with 1990s multiculturalism: the heavily debated pluri-multi model (ILDIS 1993; Albó 1994) that speaks to a line of incorporation while maintaining many of the established hierarchies between mestizo and indigenous peoples. The representation of indigeneity is a crucial selling point for Bolivian musicians touring Japan, but in Japan some of the specificity demanded at home blends into a more general need to represent regionalized cultural difference. Within Japan that world of difference is often glossed as "Andean."

In this case I explore the parameters of cultural encounters put into place by the export of culture, entering into that conceptually slippery realm between culture as a way of life, and culture as commodified representations (Wade 1999: 449). While development projects under globalization have emphasized "culture" as "a way for poor people to help themselves and the economy" (Elyachar 2002: 500), in this work, Bolivian musicians have placed Bolivian local cultures on national and transnational performance stages. For

Japanese audiences, Bolivian musicians have to perform an ethnic, indigenous, or Andean Other, but following Jan Nederveen Pieterse's critique of the "ethnic economy" (Nederveen Pieterse, this volume), the business relations between the Japanese contracting company and the Bolivian musicians are not entirely bounded by conceptual categories of ethnicity. These cultural flows are transnational in the sense that neither the Japanese company nor the touring musicians work in any official capacity within their respective nation-states. But the projects of the company and of the musicians hardly subvert national projects either. The musicians consciously take on a nationalist project in their own forms of self-representation, and the contracting company sells the Bolivian show to Japanese schools – realms of the state where Japanese national citizens are shaped. The transnational cultural flows, while constructed through cultural entrepreneurs, also work within spaces of the state. These details make it worth contemplating Aihwa Ong's call to consider how globalization processes may strengthen the state rather than weaken it (Ong 1996: 6).

In the following analysis, I draw on ten years of ethnographic research on Bolivian music, ten years of performing with Bolivian musicians, a three-month music tour of Japan in 2002 with representatives of the Bolivian ensemble (Music of the Masters), and interviews conducted in both Bolivia and Japan, with Bolivian musicians who have toured Japan, and with Japanese who have been involved in playing, contracting, or sponsoring Bolivian music.[2] During the three-month tour, I participated as a performing musician, playing 75 concerts in 88 days but, unlike my Bolivian colleagues, I received no payment from the company.[3]

Bolivian musicians in Japan fit into a cultural economy where their coveted qualities for employment include their development of multiple skills, their flexibility in terms of accepting contracts for lengthy tours, and their ability to convincingly embody cultural difference. Their employment situation mirrors that of the "cultural entrepreneur" who, according to Andrea Ellmeier, is replacing "the artist" in a post-industrial cultural economy (Ellmeier 2003: 10). The kind of flexibility required by the musicians' labor conditions sits in juxtaposition to the rigid views of culture that frame their staged performances. I argue that the late capitalist laboring in cultural difference is fundamentally driven by this juxtaposition between the necessarily deterritorialized work conditions and the ultra-territorialized conceptualization of culture as something spatially bounded and unproblematically authentic. First, I address how Bolivian music performances in Japan are shaped by somewhat fixed ideas of what is "Andean." Second, I discuss how Bolivian musicians' labor conditions in this process are characterized by repetitive tasks and flexible contracts, and how some Bolivian musicians are disturbed that they are not, under these conditions, treated as "artists." Finally, I address the symbolic work of cultural difference that operates in this particular global economy of music.

"Andean" music in Japan

Not every Bolivian musician gets invited to tour Japan. All those who have toured speak of an initial connection or contact. For Music of the Masters, that contact was ten years in the making and came through a man named Koji Hishimoto. In 1989, Hishimoto arrived in Bolivia with the intention of staying for three months and he stayed for seven years. His main purpose for being there was to learn to perform Bolivian music, specifically the wind instruments of the notched flutes and pan pipes. When Hishimoto met Rolando Encinas, a well-known notched flute player and director of Music of the Masters, in his broken Spanish, Hishimoto pointed at Encinas and said "teacher" and then pointed at himself and said "student," expressing the desire to set up a master–apprentice relation. Thus began an apprenticeship that became a partnership in the composing, arranging, and performing of Bolivian music, both within and outside of Bolivia. When Hishimoto returned to Japan, he began working as a guide and translator for *Warao Neko*, Laughing Cats – a Japanese entertainment company that contracted international groups to perform in schools and theaters throughout Japan. After a few years, Hishimoto facilitated an invitation to Encinas and the latter opted to tour under the name of his group, Music of the Masters.

Music of the Masters in Bolivia consists of a 24-to-30 person orchestra that interprets the "master" composers of Bolivian music, drawing on mestizo-Creole genres as well as indigenous music from the highlands and lowlands of Bolivia. Much of the orchestra's local appeal derives from its novel form that breaks from the five-person ensemble that, since the 1960s, was popularized both nationally and internationally. Encinas developed the ensemble's repertoire by digging through archives for original scores, talking to Bolivian composers and their surviving relatives, and conducting fieldwork in the Bolivian countryside. Most of the members of the group are mestizos from the lower middle class of Bolivia; only a few members more strongly identify with indigenous Aymara backgrounds.[4] The ensemble has never had any official connection to the state, and Encinas has maintained a fiercely independent attitude about the group's productions.

In Japan, Music of the Masters took an entirely different shape, beginning with the number of musicians who traveled. Only five musicians were invited from Bolivia, a number set by the Japanese company. If Encinas directed Music of the Masters in Bolivia, while on tour in Japan, he turned over much of the artistic direction to Hishimoto. Encinas depended on Hishimoto to musically translate a Bolivian nationalist artistic project into a show that would appeal to Japanese audiences. In a sense, this task entailed playing "Andean" music. "The Andean world," a knowledge construct that once was the realm of extensive anthropological inquiry, has since the 1990s come under extensive critique (Starn 1991). Just as scholars of a post-Edward-Said world want to avoid Orientalism (1978), scholars of the Andean region tiptoe around any possibility of committing Andeanism in their current work. Nevertheless, the power of Andeanist discourse remains strong within the shaping of transnational pro-indigenous development projects (see Andolina, Laurie, and Radcliffe forthcoming), and also within the imaginations that define transnational Bolivian music performances. To land touring contracts, Bolivian musicians perform Andeanism. While my more general fieldwork on this topic has revealed that Japanese professional musicians and hobbyists of this music manage extensive and detailed knowledge of different Andean music traditions, a general listening audience tends to lump Bolivian music under the general term "folklore." To this term the Japanese did not attach any national or ethnic adjective and understood the term "folklore" to mean *only* music from the Andes, and more specifically, any music that included the playing of *quena* (notched flute), *zampoñas* (pan pipes), and charango. In Japan, "folklore" is synonymous with "Andean music" and there is no other kind of folklore.

The Japanese talk about being drawn to this music through Simon and Garfunkel's late 1960s recording of "El condor pasa." What audiences knew as the song that began, "I'd rather be a sparrow than a snail . . ." was a 1933 composition by the Peruvian composer, Daniel Alomía Robles. Like tango in Japan, which had to go through the European metropoles before becoming popular in Japan

(Savigliano 1995), "El condor pasa," as filtered through the global song machine of the USA, was key in ushering in a European and Japanese boom in Andean music. Just as tourism demands "a generalizable culture that can be packaged and purchased" (Medina 2003: 364), the commoditizing of Andean music for a global audience implied a similar process. Simon and Garfunkel's version of "El condor pasa" did this work of translation, bridging, and generalization. No performance of Andean music in Japan was complete without some rendition of this tune, and this became the second theme played for every performance we gave. Hishimoto would introduce the group, discuss the high Andes mountains that marked the group's geographic place of origin, and talk about the condor that flies high in those mountains. As a backdrop to our musical performance, a mural of the snow-capped Illimani mountain provided the visual cues for the context of La Paz, Bolivia. Hishimoto would gesture through an imaginary map of South America and would end by mentioning the Peruvian origin of the piece we were about to play. Hishimoto explained to me that the Andes mountains and "El Condor Pasa" were points of reference on the way to situating the audience within the frame of Bolivian music proper. As with most audiences, some people learned better than others. At the close of our concerts, Hishimoto asked the students to tell him where this music was from. He often received a range of incorrect answers including "Africa," "Mexico," Brazil," and "Argentina." At hearing these answers, Encinas would shake his head in frustration: "You can play your heart out in 40 degree [Celsius] heat and they still don't know you are from Bolivia." These incorrect answers never referenced the USA or any Asian country. Even for the inattentive listener, these performances were about non-US and non-Asian Others.

After playing "El condor pasa," we moved into Bolivian themes, but some genres were seldom, if ever, played in these performances. In developing the repertoire for these shows, Hishimoto steered away from any genres that did not feature a steady one-two beat (2/4 or 4/4 meters). He commented on the desire to evoke audience participation in the form of hand clapping and in some cases, even dancing. For an audience whose members do not know the music, this kind of participation proves more difficult in genres played in more complex meters. During the Japan tour, we played no *bailecitos*, no *carnavales*, and only a few *cuecas* – all genres that form an integral part of Music of the Masters' repertoire when we perform in Bolivia, but also genres that feature combined meters of two and three (6/8 and 3/4).[5]

In his choice of repertoire, Hishimoto also emphasized a heavy drum-beat, another element that could potentially activate audience participation. Nearing the end of the tour, at about performance number 70, I had reached far beyond saturation point with the set repertoire and I begged Hishimoto to substitute just one piece, any piece. I was desperate; I suggested a different tune in one of the genres already included in the program. Music of the Masters had a vast repertoire – about 80 compositions at the time – but our Japanese performances varied only slightly from one concert to the next. Hishimoto responded by sighing quietly, shaking his head, and telling me that the piece I suggested did not include a part for the *bombo* – the large goat-skin-covered drum. My suggestion was quietly rejected as Hishimoto followed the aesthetic of keeping the sonorous element of the drum. I mustered up a smile and feigned enthusiasm for performance number 71.

Our performances were intentionally educational and included several interludes of explanatory discourse. A part of every concert included a "demonstration of instruments," a section through which Hishimoto would talk as the appointed musician would solo on the instruments perceived to be quintessentially "Andean." A great number of these spoken interludes remained in our shows, even when we played for regular audiences in a theater. When I attended concerts of other Japanese musicians who played "folklore," their performances were also peppered with all kinds of explanations. *Talk about the music* seemed to be an integral part of musical shows that were presented as culturally different. Unlike more experimental approaches that might present culturally different performances with no explanatory discourse, expecting the audience to approach this difference as an audience might approach avant-garde works (see Kirshenblatt-Gimblett 1998: 205–41), the cultural difference of "Andean music" in Japan was assumed to need ample explanation. Cultural difference was not something to simply experience as a good time, but something to be contemplated educationally and pedagogically. "Bolivian music" in Japan was shaped to meet audience expectations about "Andean music" or folklore, and these expectations emerged through a global music economy that treated culture as difference. Japanese preconceptions about Andean music, perhaps set in the 1970s, continued to fix performance parameters for touring musicians of the early twenty-first century.

Laboring for Laughing Cats

While interest in an essentialized cultural difference helped generate the musicians' contract for a tour of Japan, the musicians' labor conditions mirrored the widespread tendencies towards labor flexibility and flexible specialization that characterize post-Fordist capitalism (Amin 1994: 4). Capitalism has a long-term trajectory of "mobilizing and homogenizing the labor force on the one hand and formalizing difference on the other," and the contemporary focus on autochthony is a continuation of "the paradoxes of capitalist labor history" (Geschiere and Nyamnjoh: 2000: 427, 449). As Arjun Appadurai has suggested, "globalization of culture is not the same as its homogenization" and ultimately these global processes are articulated through the twin politics of sameness and difference (Appadurai 2002 [1996]: 57–8). Let me turn to more specific details of the tour that show how these paradoxes of capitalism work on the ground.

The selection of five Bolivian musicians for the Japanese tour was a delicate matter and one that stirred jealousy on the home front. As Aihwa Ong points out, transnational flows of culture and capital have not brought an accompanying equality in the mobility of subjects (Ong 1999: 11). In general, this ensemble's musicians are the disempowered non-mobile subjects who have not been emancipated by global flows. Their mobility has been limited to very specific, short-term tours. Those who can participate in this temporary transnational cultural labor usually are underemployed at home and already existing within flexible labor contracts. If they are not in this tenuous position in Bolivia, a three-month tour and absence from any more stable employment is likely to move them into the realm of labor flexibility. When Saskia Sassen discusses the classifications for foreign workers in Japan, she names the first category as including "artists, religious personnel, and journalists," and she states that these categories "do not represent the typical migrant worker" (Sassen 1998: 61). Bolivian musicians in Japan held entertainer or artist visas, occupying this atypical migrant worker position, but they often categorized their own labor activities as "service." At home and abroad, I have heard Bolivian musicians discuss their work in terms of "cultural service." While some musicians made this categorization with an attitude of "pride in honest work," other musicians complained about touring conditions and not being "treated like an artist." While Bolivian musicians may have entered as atypical migrant workers (i.e. as "artists") they did not always feel treated as such. In selecting those who traveled, Encinas considered the musical demands of a five-person ensemble as well as the intense personal demands of working and living in shared space over a three-month period. Controversy emerged at home because, for the guaranteed earnings of this work, just about everyone in the ensemble would be willing to drop everything in Bolivia and tour Japan. In Japan, Hishimoto always joined the group, although initially, much like my own situation, he had to persuade the company to permit his embodied Japanese presence in the middle of this Bolivian performance spectacle.

In a sense, Laughing Cats purchased the labor of the musicians for a set number of days and during those days the company would schedule as many performances as possible. Laughing Cats also arranged for the musicians' visas, paid their airfare and hotel, and distributed a US$30 per diem to each touring musician for the purchase of food. Upon finalizing the tour, sometimes a mere 36 hours before departure, each member of the group received from the company payment for their work. Operating out of a small office and rehearsal space in Mitaka, Tokyo, Laughing Cats would often have two or three groups simultaneously on tour; a Chinese theater group and a Korean dance troupe were all on tour when I was in Japan in 2002. Laughing Cats also sent Japanese troupes on performance tours abroad. For example, Hishimoto was once contracted to accompany a Japanese theater troupe in a tour of Mexico and Costa Rica. Hishimoto's work, as well as that of the tour manager, and sound engineer were – like the musicians' arrangement – short-term labor contracts that lasted as long as the tour, usually two or three months of work. Flexible labor contracts shaped the work of both Japanese and Bolivian members of the tour.

Among Bolivian musicians, a Laughing Cats tour is viewed as a no-frills arrangement: lots of hard work and, according to some musicians, not a fair remuneration. Everyone on a Laughing Cats tour – musicians, manager, and sound engineer – was involved in the grunt work of setting up and taking down a performance, something that would happen one to three times a day, depending on how many performances were scheduled for the day. A particularly heavy workload would include up to three performances at three different schools within a single day, and this pattern was often repeated over several days within a single week. Tour members stayed in business hotels in the cities and in *ryokans* or traditional hotels in the countryside; these arrangements often involved sharing rooms with other members of the tour. We all traveled in a medium-sized bus/truck, with just enough seats for those on the tour and ample space behind the seating area for sound equipment, instruments, costumes, and suitcases. After finishing the performances for the day, the group would often have three to eight hours of ground travel before arriving at the destination of the next days' performances and the night's accommodations. While some Bolivian musicians, because of the working conditions, traveled once with Laughing Cats and never returned again, other groups continued to return for annual tours. In spite of these working conditions, a contract with Laughing Cats still represented remunerated employment in music that simply could not be matched within Bolivia.

Hishimoto told me he once entered a school director's office and on his desk he saw over 100 brochures advertising different performing groups. Many companies were competing for these contracts with schools. With this kind of competition, company workers, even those on short contracts like Hishimoto, our manager, and our sound engineer, felt the pressure to meet high standards of work. Those standards included a meticulous attention to punctuality, a serious attitude towards work, and the attempts to meet perceived audience expectations in a performance of cultural difference. Punctuality was key to the success of the company's work. Our Laughing Cats manager would drive the bus into the school grounds about 15 minutes before our scheduled arrival. But to make that first appointment of the day, we had usually traveled five or six hours the previous day, making sure we were at a hotel within 20-minutes' drive of our first performance site. Each day the manager carefully studied maps, because this job was not about showing up on time to the same place of work every day, but rather about punctually arriving every day at two to four different, and previously unknown, places of work. Ideally, we would arrive two hours before show time. Within this timeframe, we had ample time to unload, set up, do a sound check, change our clothes, and perhaps even rehearse problem spots of the performance. Timing was crucial in the school context because students were scheduled to attend the special assembly of our performance, and these performances had to fit within the already highly regimented school day. If we gave one performance during the day, being punctual seemed quite easy, but when we had three performances in a single day, at three different schools, being punctual became a real challenge indeed. In one case, we literally ran to set up a second show for which we had exactly 25 minutes from arrival to show time.

Each day involved a serious attention to this kind of work until the final show of the day. The Bolivians emphasized the importance of teamwork to unload, set-up, and strike the set. As the tour progressed, internal critiques inevitably surfaced about someone who wasn't quite doing their share. We usually had some time between setting up and doing the sound check. I used this time to play scales and exercises on the violin. As we moved into the fall months, and the mornings became cooler, many of the Bolivians began to grab a game of basketball during this short interlude. Most of our school performances took place in gymnasiums that doubled as theaters for staged events, and usually a basketball could be found in some corner of the venue. As Jhuliano Encinas, the guitar player told me, the physical activity warmed them up and helped them bring renewed energy and enthusiasm to what could easily become a monotonous performance routine. For the Bolivians, this kind of enthusiasm and energy, even if feigned, was a necessary element of good work on the tour. In fact, I was constantly under criticism by other group members because I apparently did not meet these standards of visible and demonstrable enthusiasm. In the middle of the show, another musician would annoyingly remind me to smile. I thoroughly enjoy performing with these musicians – the fatigue of the tour aside – but while my stage presence in Bolivian performance contexts might be critiqued for my musical interpretations, in those contexts I was never critiqued for lack of smiles or enthusiasm. Exaggerated enthusiasm was a particular labor demand in the Japanese transnational context. Musicians who had previously toured with this company would express their dissatisfaction in terms of whether or not they were being treated "like an artist." Being "treated like an artist" translated to accommodations in five-star hotels, comfortable travel arrangements that included air travel and bullet trains for long distances, and the ability simply to arrive at a performance, pick up an instrument, and play. Making music on these tours leaves little room for an artistic persona. The work is something of a mix between service, of teaching young Japanese students about Andean music, and assembly line production of aesthetic experiences.

On occasions, the Bolivians' attention to pumping up energy and enthusiasm clashed with some of the Japanese's expectations about work. For example, one school director became very angry when she saw the Bolivian musicians playing basketball. The Bolivians thought it was because they had not asked for permission to do so. According to Hishimoto, that was not the issue; rather, the director was upset because the school was paying them to work as musicians and to put on a show. She could not believe that these paid musicians were playing basketball on her yen, so to speak. In the context of this incident J. Encinas defended their game of basketball as a way to bring energy to their performances. On a daily basis, the musicians sought ways to laugh and joke on stage, and as the tour progressed I saw this as another strategy for keeping up enthusiasm. For example, in the second half of the show, we always introduced ourselves in Japanese – one of the few lines of Japanese we all had to learn. No one in the audience really knew our names, so we made a game out of giving ourselves silly names. Our sound engineer, who was studying Spanish and who understood more and more about the daily on-stage jokes, would complain about our lack of a serious attitude. I would often catch him hiding a smile as he crouched behind the sound board and tried to keep a straight face.

Working in the culture mines entails serious attention to playing a role on stage: the role of musicians who enthusiastically want to share the music from their country. Tour members commented on the fact that this kind of energy and enthusiasm, through one performance after another, is probably the most difficult part of this work. Bolivian musicians' strategies for fulfilling these roles did not always meet with the approval of Japanese hosts who sometimes held conflicting ideas about how one should approach working in music and cultural difference. Bolivians were expected to be serious about work, but on stage they were expected to embody cheerful, happy-go-lucky natives who showed not an inkling of being worn down by the repetitive tasks of the tour.

After a three-month tour, musicians returned home with money for everyday expenses, but also with limited capital that was invested in the construction of a house for one's parents, in sound equipment that could be "made to work" as another source of income, in the purchase of a car that could be "made to work" as a taxi, and in other living arrangements that moved one out of what is a very precarious position for a Bolivian musician – the payment of monthly rent.[6] Money earned while doing cultural work in Japan expanded one's repertoire of economic survival strategies at home. A Laughing Cats tour involved hard work, but work that Bolivians, coming from a country where remunerated employment is so scarce, were more than willing to endure.

To participate in these short-term employment "opportunities," Bolivian musicians have to be already within flexible labor conditions at home. While the work in Japan provides relatively significant income on a temporary and unpredictable basis, some musicians explicitly remarked on how the treatment on these tours leaves them feeling quite far from the heightened pedestal of "artist." In the next section, I turn to the symbolic work of performing cultural difference in the Japanese national context.

Transnational politics of cultural difference

In its labor contracts, the company preferred musicians from what might be called "Third World" contexts. I would suggest that the reasons for these choices were connected to tight competition for shows, economic limitations, and Japanese ideologies about national homogeneity. Bolivian "culture" is mobilized as a difference and is set up to work with the official ideologies of Japan as a homogeneous nation (Kelly 1991: 413; Masden 1997: 56; Sassen 1998: 59). The myth of Japanese homogeneity, based on the metaphorical associations of "Japanese blood," has come under increasing strain within the contexts of Japanese returnees and foreign workers in Japan (Yoshino 1997). Pluralization of Japanese society is in fact challenging the long-standing myth of a homogeneous Japan (Murphy-Shigematsu 2000: 215). According to Kirk Masden, in the face of this pluralization, many of the official policies of the Japanese Ministry of Education continue to uphold the myth of homogeneity (1997: 29). Bolivian music, transformed into Andean music, works as a marker of cultural difference within the Japanese context, but it does so without threatening this country's core ideology of homogeneity.

Partway through the 2002 tour, my Bolivian colleagues had to complete immigration paperwork, to extend by three days their stay in Japan under the "entertainers" visa. A company representative accompanied the group to complete their paperwork and en route to these bureaucratic dealings, I had a lengthy discussion with her. From Latin America, she said they hired groups from Mexico and Bolivia. She told me that Japanese like "happy music" and she quickly categorized under this term, the music from these two countries. She told me that the company had tried to hire a Native American group, but they were unable to do so because the performing group demanded fees at a level beyond the company's budget. Although I did not ask her directly about contracting groups from the USA, she volunteered her thoughts, responding to me as if I might wonder about my own country's representation in this company's work. The company had considered bringing a gospel group, or even a "country" music band, she said. But they were afraid the fees would be too high, particularly since they would be coming from the USA. The geopolitical asymmetries of culture and power seemed to predetermine who would be hired as a performance spectacle for Japanese school children. While some Bolivian musicians, on the basis of tour conditions alone, would not tour a second time with Laughing Cats, there were still more Bolivian musicians who would jump at this employment opportunity. Bolivian and Mexican groups would accept the company's fee while Native Americans, from the US national context, had set a fee above what the company could pay and still run as a business.

The company representative also mentioned that performances from Asian countries were high in demand. Since the co-hosting of the World Cup, she suggested, the Japanese had showed a renewed interest in Korea. She said, "There is a feeling we are becoming friends with Korea." The head of the company is married to a Chinese woman who also works in the company. Some of the interest in "Asia" may be facilitated by this personal tie to another Asian country, but it may also be part of Japan's "return to Asia" – a renewed interest, since the 1990s, in the rising economic power of other Asian states (Iwabuchi 2002b: 547). According to Hishimoto, primary and secondary schools are required to sponsor one special performance each year, but it does not have to be a foreign group's performance. The Laughing Cats' representative told me these foreign shows were strongly recommended by the Ministry of Education. In qualifying her explanation, she referred to Japan's geography, saying that as a country of islands, it is difficult for Japanese to get to know other cultures. From her perspective, these foreign spectacles helped break an imagined cultural isolation in Japan. According to the company representative, these cultural performances were perceived as a way for Japanese students to get to know the world. Even as the space-time compressions of labor and symbolic capital are surely at work here (Harvey 1989), narratives of perceived Japanese isolation and homogeneity seem to propel these cultural performances in the first instance. Economic conditions determine who the Japanese company invites to occupy the position of exotic other. Cultural workers from First World countries, even if relatively disadvantaged within their own national contexts, can afford to outprice a possible contract for touring Japan. Many cultural workers from Third World contexts are not in an economic position to be so choosy.

Laboring in Bolivian culture within the Japanese context brings up old but still pertinent issues about culture and power. In his 1989 essay, "Notes on the Global Ecumene," Ulf Hannerz commented on Japan's "low cultural profile" and questioned a simple reading of political and economic centers as exporters of culture. Hannerz takes apart some of the assumed patterns of cultural flows that supposedly occur between centers and peripheries, and argues that not every wealthy nation follows the US example by exporting its "culture" (Hannerz 2002: 38). Hannerz's "culture" within this argument refers to mass-mediated popular culture. In the same year, Renato Rosaldo published an argument about an altogether different angle on "culture." Rosaldo argued that those who seem to have culture (i.e. seem different) have less power and those who seem not to have culture have more power (Rosaldo 1989: 196–217). Things have shifted since these two arguments were published. For example, Japan's economic bubble has burst and scholars are noting Japan's current export of its mass-mediated popular culture (Iwabuchi 2002a: 4–5), particularly within other Asian countries (Ching 2001: 294). In relation to culture as perceived difference, Rosaldo's ratios still seem to operate. Within Japan, Japanese may view Bolivians as "having culture" even though the latter may lack the economic resources that would alter fundamentally their standard of living. But the Japanese see themselves as "having culture" too. Rosaldo's ratios are suddenly taken for a spin. When asked to compare their lives with those of Bolivians, the Japanese I interviewed reflected on their own fast-paced lives and expectations of an ultra-serious attitude, while turning to Bolivian cultural performances to lighten up. While Bolivians and Japanese recognized how they occupied an exotic slot, the socioeconomic inequalities between them still made their mutual encounter into a hierarchical one.

Bolivia has long been a country whose natural resources have been mined to the advantage of non-Bolivians and a minority elite within Bolivia. Silver was mined for the Spanish empire, while tin benefited a few mining Bolivian barons. Gas has become the next item of extraction. In October 2003, popular sectors put the brakes on one more extraction for foreign benefit; they forced President Gonzalo

Sánchez de Lozada to resign and called for a national referendum on the sale of gas. Since 1985 and the Supreme Decree 21060, Bolivian governments have applied neo-liberal economic packages and walked a precarious path between the pressures of external sources of aid, like the USA, and the majority of Bolivians who have yet to experience any benefits after almost 20 years of structural adjustment. Alongside structural adjustment came an eventual rise in indigenous politics, both on the ground and within some government initiatives of the early 1990s. In the 1990s, non-governmental organizations (NGOs), the government, and indigenous groups played roles in the staging of indigeneity for international funding agendas. In using the performance metaphor, I do not imply that these activities were less significant or that they made indigenous peoples "less real": this would be to read the theatrical metaphor through Western ethnocentricities that view acting in terms of falsehood (Schieffelin 1998). Rather, I want to emphasize that these representations of indigeneity, although perhaps inclusive of more people who consider themselves indigenous, are like musical representations of indigeneity in that they are articulated through a multiplicity of groups and interests.

On the national musical stage, Bolivian political transformations were transcribed into the previously mentioned attention to specific indigenous performance practices. In the midst of Bolivia's dire economic straits, the country attracts tourists to see its stunning lakeside views, appreciate its snow-capped peaks, visit its Pre-Columbian archeological sites, observe Bolivians dancing in colorful fiestas, and hear its diverse range of musical styles. While economically strapped, Bolivians present a visual pageant rich in cultural difference, and that cultural difference has been exploited for tourism and extracted for world export.

One might write of the relative power of the Japanese next to the relative powerlessness of the Bolivians, and the importation of "culture" from the less powerful context resonates with Rosaldo's relative positionings of culture and power. But this case shows that both the Japanese and Bolivian workers who labor in this deployment of culture operate under similarly tenuous or "flexible" labor conditions; the "cultural entrepreneurs" (Ellmeier 2003) here are Japanese as well as Bolivian. While many discussions about power and cultural difference often hold implicit a Western/non-Western dichotomy, a consideration of Bolivian music in Japan calls forth alternatives that do not necessarily fit this mode. The meaning of culturally performed difference in Japan may not fit exactly that pervasive monolithic dualism, and to understand the work of Bolivian music in Japan, one must consider how difference operates within – or perhaps more appropriately said, outside of – ideologies of Japanese nationalism.

Conclusion: the fixed and the flexible

In this chapter I have explored what it means to labor in the production of Bolivian music performances in Japanese schools. The terms of that labor are shaped by what the Japanese audiences expect to get from a performance of Bolivian music. Cultural difference, I argue, is the Japanese's principal desired product of this labor, and the staged cultural differences of Bolivians symbolically work to sustain a Japanese nationalist ideology of homogeneity against a heterogeneous non-Japanese world. Touring Japan in this way is not generally the labor of love about which Bolivian musicians dream. There is the occasional performance in a spectacular Japanese theater, where the combined elements of a state-of-the-art sound system, lighting display, supporting crew, and crucial changes to the performed repertoire make for an incredible performance. But that is one performance in twenty. Encinas often reiterated that we were involved in a cultural labor (*un labor cultural*). For many touring members this labor made other economic strategies at home financially possible. Making money in the Japanese culture mines also permitted many of these musicians to engage in creative musical composition and performance in other contexts. Laboring in culture in Japan made it possible to engage, at other moments, in music as a labor of love. Bolivian musicians in Japan articulate complex positions in relation to their work. At the level of the daily repetition of tasks, they admitted the straightforward economic incentive in touring and talked about small capital investments that they hoped might tilt their economic position at home away from precariousness. On the other hand, these musicians are extremely proud of their country's music; they want it to be valued beyond the local subway station, and they discuss the tours of Japanese schools in terms of building future audiences. Celebration of their own music in Japan works like other celebratory narratives of world music that "tend to normalize and naturalize globalization" (Feld 2001: 197). Moreover, celebration continues despite the problematic side of globalization, which leaves culture workers with deep nationalist sentiments even as their decontextualized labor builds "distantiation from place and its sociomoral pressures" (Comaroff and Comaroff 2000: 303). In the face of the shrinking Bolivian nation-state (see Gill 2000), sentiments about national culture continue to fuel these transnational artistic activities.

While the performance expectations for Bolivian musicians in Japan can be quite rigid, their labor contracts are fluid and unfixed. A three-month tour may be repeated over several years, but no guarantee is ever made for future contracts. After four or five years of touring, some Bolivian musicians have felt the results of being temporarily or perhaps permanently left outside of these contracts. These conditions divide musicians in the Bolivian context and push workers into individually negotiated entrepreneurial ties. While these conditions are perhaps not unique for people working in music, they are exacerbated by the effects of failed neo-liberal reforms that may have been dressed in the clothes of multicultural politics, but that have not brought acceptable living conditions for the majority of the Bolivian population.

When Bolivian musicians tour Japan with Laughing Cats, both Bolivians and Japanese are involved in the work that produces and repeats cultural difference in a staged form, and those closest to the tour activities, Japanese and Bolivians alike are limited by the same short-term "flexible" terms of labor. Globalization of labor has harnessed culture workers at home with culture workers from abroad to produce, repeat, and fix cultural difference. In this process, ideologies of Japanese nationalism are fortified and Bolivians experience a sense of national pride in what they are doing, in spite of the fact that they are often first seen as Third World citizens, Andeans and Others before they are considered Bolivians.

Notes

1   In another project, I have found that in Bolivia, indigenous music begins to emerge as staged "folklore" as early as the 1930s (Bigenho in press).

2   My earlier research was supported by Fulbright IIE 1993–4, Fulbright Hays 1994–5, and Hampshire College Summer Faculty Development funds 1999–2004. Hampshire College sabbatical leave in 2002 made possible my presence on the music tour in 2002. A Whiting Foundation grant made possible my research in Japan and Bolivia during the summer of 2003.

3   With my sabbatical salary from Hampshire College, I paid my own expenses in Japan, a financial position that somewhat puzzled the Bolivians. I had played

with these musicians over the last ten years, and they were happy to have me on the tour, but they wondered how I could receive my salary when I wasn't working at my institution. I explained that I was conducting research. They nodded and got on with what for them was the real work of the tour, and in that real work, I participated as one more team member.

4   Over the last ten years of performing with Music of the Masters, I have noted a major generational shift in the membership of the orchestra. Whereas the original membership consisted of many people who were already working and/or heads of households, the current membership is weighted more heavily with young people who do not necessarily maintain or contribute to households. The ensemble has never been pushed in a commercial sense, and some of the older members had to give priority to other groups through which they earned more money in their music performances.

5   In other work, I have written about the performance practices and disputes about playing these meters in Bolivian music (Bigenho 2002: 47–54).

6   Within Bolivia, a peculiar relation to property exists called "anticrético." The owner of a house or apartment leases the space to a person who is willing to put down a sizeable sum of money. The person may occupy the space for a year or more without paying a monthly fee in rent. Upon leaving the owner must return the lump sum to the "tenant." When musicians do not have the means to purchase a home or apartment, they use anticrético as a step out of monthly rent.

6 Social capital and migration – beyond ethnic economies

Jan Nederveen Pieterse

How does social capital relate to cultural difference? Considering the importance of cross-cultural trade and economic relations, historically and now (cf. Griffin, 1996, 2000), one would expect this to be a salient issue, but it hardly figures in the literature. The conventional assumption is that social capital is culturally bounded. In most literature this is precisely taken as the strength (particularist loyalties, lower transaction costs, and so on) and the weakness of social capital (group exclusiveness). There are two major strands in the literature. In one, cultural difference fades into the background and informal social relations and group bonds are at the foreground; this is the course taken in the work of Coleman and Putnam. In the other strand, culture (usually reified as "ethnicity") is both a resource and boundary of social capital; the latter terrain is the focus here. In a sense, this line of enquiry appears as an extended commentary on "ethnicity" as the pattern of a particular type of social relations, much of which is modelled in turn on the role of the Jews in commerce. This was the subject of classic studies in the field: Simmel's essay on *The Stranger* and Sombart's sequel study of *The Jews in Modern Capitalism*. An implication of these studies is that ethnic social capital is a pre-modern hangover in modern times.

Immigrant enterprise is now widely considered to be a factor in the economic dynamism of many countries. A matter of keen debate in the USA and Canada is whether immigrant enterprises are more significant employment creators than domestic enterprises. Headlines such as "Millions of Immigrants Needed to Sustain Economies" (Wordsworth 2000) are increasingly common in Canada, Germany, Italy, and several other countries.[1] Part of the wider backdrop is graying labor markets in several OECD countries. In addition, specific immigrant groups are viewed as making special contributions, such as Indian software programmers and Chinese engineers and programmers in Silicon Valley, and are actively sought after. While countries are relaxing rules to facilitate the faster deportation of illegal immigrants, they are relaxing immigration laws to facilitate bringing in desired migrant entrepreneurs, particularly with a view to attracting dot.com enterprises and programmers.[2]

In this context, several stereotypes of immigrant enterprise are gradually being left behind. For example, a study of Tunisian immigrants in France shows that, unlike in the 1970s, immigrants are now more often self-employed, community ethics give way to economic rationality, and commercial organization and transnational networks are developing. Research suggests that this also applies to Asian and Turkish immigrants in France (Boubakri 1999), and there are similar findings in Germany (Özcan and Seifert 2000).

Meanwhile, in most research, attention remains focused on the *ethnic* character of enterprise. This chapter argues that "ethnic economy" is more often a misnomer than accurate. Cultural capital matters alongside social capital, but viewing it as "ethnic" in character is not helpful and is likely to be misleading. The second general point is that immigrant economies are often embedded within cross-cultural economies. The chapter considers whether "immigrant economy" would be a more insightful terminology, but finds similar problems.

In relation to social capital, a key distinction runs between *causes* of social capital (norms and values, or "habits of the heart," and social networks) and *outcomes*, such as lower transaction costs (Newton 1999). Among causes, a distinction runs between strong and weak social ties. Further distinctions run between *bonding* social capital (strong ties among close relations), *bridging* social capital (weak ties among people from diverse backgrounds but of similar socioeconomic status), and *linking* social capital (or "friends in high places"). Considerations of cultural difference or "ethnicity" apply across these different dimensions of social capital and take on cultural hues, that is, each apply within and across cultural settings. The question of cultural difference and social capital arises in three different contexts: immigration and migration, transnational enterprise, and ethnically diverse societies. In this treatment, the emphasis is on migration and immigrant enterprise.

A related question is how is cultural difference conceptualized? Is "ethnicity" adequate or burdened by time? (for details, see Nederveen Pieterse 1997a, 2002). Much literature and reporting on ethnicity is fraught with friction, tension, antagonism. The media report on ethnicity mainly when it generates problems, oppression or conflict, in line with the media principle "when it bleeds it leads." But what of the situations when ethnicity does not entail conflict or when conflict is minor? This chapter opens by probing the notion of social capital, while the next section deals with the problematization of "ethnic economies" and "ethnicity." This leads to shifting the focus onto cross-cultural enterprise. Because a historical dimension is often missing in this line of research and focusing on the present confines analysis, immigrant economies are considered here in a historical setting. This yields several types of cross-cultural economies, which can be linked to varieties of social capital. The closing section considers the policy ramifications of the shift in orientation from ethnic to cross-cultural enterprise.

Social capital

The theme of social capital emerges on the heels of human capital. Just when the importance of capabilities, capacitation, and enablement is recognized, the attention shifts – "It's not what you know, it's who you know!" (Barr and Toye 2000). Also on the horizon is cultural capital, and another newcomer is natural capital. A background consideration (discussed below) is that none of these would add up to much without economic resources. And so the debate runs the course of several forms of capital – economic, physical, financial, human, cultural, social, political, natural – and eventually comes back, full circle, to economic capital.

Social capital is usually defined as the capacity of individuals to gain access to scarce resources by virtue of their membership of social networks or institutions. Putnam gives a wider definition of social capital as "features of social organization such as networks, norms and social trust that facilitate coordination and cooperation for mutual benefit" (Putnam 1993: 67). Social capital is a notion of the times, "the latest conceptual fad across the social sciences" (Fine 2001a). A hybrid notion, social capital mixes angles and approaches that used to be wide apart. It brings the "social" into economics and, by the same token, looks at the social from an economic point of view. Its social angle on the market comes at the price of a market angle on the social. One is not sure whether just to scratch one's head or pull out one's hair. The appeal of social capital can be read both as an agenda in its own right and as a sign of the times.[3] The significance and appeal of social capital are that it serves as a bridge between sociology, economics, and politics, as a linking concept that bridges diverse fields and invites interdisciplinary research. In the process, it presents ample problems.

World Bank language refers to social capital as "the glue that holds society together." Social, all right – but capital? This is a very particular way of looking at social relations. The "social" of course figures in many approaches such as network analysis in anthropology, social distance in sociology, widening chains of interdependence in configuration sociology; and reciprocity and trust, solidarity and belonging are other ways of looking at social relations. The terminology itself is heavy baggage. The backdrop of capitalism becomes the foreground in that social capital refers to social relations and institutions that are viewed as instrumental within a capitalist framework; thus social relations and networks become "capital," assets that can be employed for income generation. For Bourdieu (1976), this was part of a problematic and social technology of domination, another glance at how the elite run things and a French equivalent of the "old boys' network."[4] With Coleman (1988), it is part of a rational choice approach to collective action and a functionalist perspective in which social relations are redefined as exchange relations. Robert Putnam (1993) establishes a link between social capital, civic democracy, and public and economic performance. In the wake of Putnam's work, linking social capital and democracy has become a well-established theme (for example, van Deth, Maraffi, Newton, and Whiteley 1999). It suggests a causal link between social connectedness → social trust → civic engagement → civic democracy (Rose, Mishler, and Haerpfer 1997: 87) and a further link to economic prosperity.

"Capital" in human, cultural, social, and natural capital holds a promise of measurability, which is a highly strategic attribute in market-driven times. Rational choice and functionalism contribute to an analytical approach that can be readily transformed into a policy package. No wonder that for some time social capital has been à la mode and in the spotlight of funding agencies. Yet social capital is a slippery concept that ranges from cultural attitudes and social practices to public policy, politics and economic development. Attending a World Bank conference on social capital, Desmond McNeill (2003) jotted down the following stray remarks: "Social capital is a battering ram to get social issues into development"; but according to an economist, "This is pure smoke"; alternatively, it is "Anthropological wine in economic bottles." These sprawling observations illustrate the perplexity surrounding the concept.

What is at stake is that social capital would make it possible to link concerns such as civil society (along with social cohesion and participation), democracy, and good governance with economic growth and development. A booming literature, particularly in economics and political science, scans the contours of social capital to examine whether it meets the requirements of clear definition, measurability, and applicability and can serve as an instrument of analysis and policy. Much current literature is concerned with conceptual clarification and is of a modelling nature, like rival exercises in reductionism. The objective is to uncover and next to instrumentalize social capital as the newest variable of productivity and development policy: "If you can't count it, it doesn't count." Whatever can be turned into an "indicator" is welcome in an age of managerialism. For now, we bracket this problematic and turn to the question of how cultural difference is conceptualized in this setting.

Ethnic economies?

The common point of departure and widely used is the notion of *ethnic economies* (for example, Light and Bonacich 1988; Waldinger, Aldrich, and Ward 1990; Waldinger 1995; Haberfellner and Böse 1999; Light and Gold 1999; Haberfellner 2000; Schmidt 2000). "An ethnic economy consists of the ethnic self-employed and employers, and their co-ethnic employees ... An ethnic economy exists whenever any immigrant or ethnic minority maintains a private economic sector in which it has a controlling ownership stake" (Light and Karageorgis, 1994: 647, 648). The assumption is that particularistic loyalties involve as well as engender trust and thus lower transaction costs. Jewish diamond traders in New York and Antwerp, passing one another diamonds for inspection on trust and without written contracts, are classic examples. The general argument is that social control is greater and the enforcement cost of non-compliance with business expectations is lower within ethnic settings.

At this point, let us pause and consider the term "ethnicity." "Ethnicity" in "ethnic economy" performs a double duty; it is defined as a social science concept (as above), while at the same time borrowing the aura of "ethnicity" from general usage. One problem is that these two uses (the definition and the image of ethnicity) cannot be kept neatly apart. What precisely is *ethnic* in ethnic neighborhood, ethnic food, ethnic economy? The term ethnicity expresses a relationship; it denotes foreignness, but a particular kind of foreignness. It seems that some national origins are foreign while others are also ethnic. Ethnicity denotes difference and cultural distance from the mainstream. So some nationalities are more ethnic than others.

In the USA, the language is ambiguous. The idea of "white ethnics" has gained currency (in the slipstream of "ethnic chic"), yet Canadians, British, Australians, or Germans are not often referred to as "ethnic." In the USA, German, British, Irish, or Scandinavian food may be foreign, but not necessarily ethnic, presumably because these cultural influences were integrated into the mainstream in an earlier phase. Yet this also applies to native and African Americans, although these are still regarded as "ethnic." Ethnicity is a marker of cultural distance, but not every cultural distance qualifies. A country's or a people's location in the hierarchy of power also matters. In the USA, Dominicans, Salvadorans, Cubans, Koreans, Ethiopians, and so on are considered as "ethnicities," but if we look more closely, this refers not to ethnicity but to *national origin*. This means that "ethnicity" serves as a descriptor of a relationship between cultures, a parameter of cultural distance and difference, which does not necessarily tell us much about the group itself. Within the host country, the nationality may be viewed as an ethnic group or minority, but this does not necessarily match relations within the country of origin. *Within* each of the nationalities mentioned there are multiple cultural groups or subcultures, identified by region, religion, sect, language, which in some contexts are called ethnic groups. In Manchester, England, Pakistanis are viewed as an ethnic group although they hail from different regions in Pakistan (Werbner 2001). Cubans from Cuba are quite different from the "Miami Cubans" (Portes 1987); among the latter, Cuban Jews are different again.

Let us consider the example of Ghanaians. Ethnic groups among Ghanaians include Asante, Fante, Brong, Kwahu, Adansi (which belong to the broader Akan ethnic group), Ga, Adangbe, Ewe, and Dagban. These groups are reproduced in the migrant communities overseas where usually ethnic associations have been formed. Ethnic associations "provide social and moral support, especially in time of bereavement and fatal sickness, much more than any tangible economic and social value for the establishment of business, hence their limited membership base and level of participation" (Amponsem 1996: 161–2). Ethnic associations exist alongside other networks such as national associations, old boys' networks, professional associations, class-based networks, women's clubs, and church networks. While there are some ethnic patterns and clusters in migration (1996: 161), migrant communities are ethnically mixed (compare Owusu 2000). Accordingly, ethnic association is only a narrow basis for social capital. As Amponsem comments:

> Due to intense competition, playing one's membership of a particular ethnic group too high leads to the risk of business being branded as an "ethnic shop" by the immigrant community at large and, therefore, risks exclusion and patronage from other members in the community. Rather, membership or affiliation with Ghanaian (national) associations such as [the] Ghanaian National Association of Hamburg, Sikaman Association in Amsterdam, and Association of Ghanaians in Toronto, even though difficult to organize in bigger cities like London and New York, are more neutral and preferable to ethnic ones.
>
> (1996: 162–3)

Thus, what from a Canadian, American, German, or Dutch point of view is an "ethnic shop," is precisely *not* an ethnic shop from a Ghanaian point of view. It would not make sense as, and could not afford to be, an ethnic shop. By labeling it thus and assuming social capital to be based on "ethnicity," we have precisely missed the point.

This is probably generally valid. Lebanese businessmen in West Africa, North America, or Australia hail from different regions and denominations within Lebanon. Although they belong to a different nationality than the host country, their social and economic cooperation need not be among Lebanese and is still less likely to have an "ethnic" basis (compare Hourani and Shehadi 1993). While Jews are generally considered an "ethnic group," Israel now ranks as a multicultural society. Korean grocery stores in California rank as ethnic shops in the media and literature, but their social cooperation need not be on an ethnic basis since they probably come from different regions in Korea. Only occasionally are we told of a *regional* or specific identification besides the national one, for example when we are informed that Palestinians owning stores in the San Francisco bay area are mostly Christians from the Ramallah area (Kotkin 1992: 236). Similarly, Iranians in Los Angeles break down into at least four different groups: Jews, Bahais, Muslims, and Armenians. This has been interpreted as "internal ethnicity in the ethnic economy"; thus, what seems to be an ethnic economy upon closer inspection turns out to be four ethnic economies (Light, Sabagh, Bozorgmehr, and Der-Martirosian 1993, 1994).

The foundations of ethnicity may include region, alleged common descent, religion, or language, and these can also intersect one another (same language, different religion, and so on). Take the case of religion. Instances where religion and ethnicity (i.e. region, language, alleged common descent) coincide, such as Sikhs, Parsis, Ismailis, and Jews, are rare by comparison to cross-cultural religions. Besides, these groups are not homogeneous either: not all Sikhs, Parsis, Ismailis, or Jews follow their religion. The "world religions" are typically cross-cultural and so are their adherents, in their countries of origin as well as in countries of immigration. A Shiva temple in India or Nepal may well be a meeting place for Saivaite Hindus from different parts of India or Nepal. The 400 Hindu temples that exist in the USA have typically been built by Hindus from different regions of India, such as Gujaratis, Bengalis, and Tamils (Tambiah 2000: 181). The metropolitan area of Phoenix, Arizona counts six mosques that are places of worship for Saudis, Sudanese, Pakistanis, Lebanese, Maghribians, and so on. The largest mosque is located in Tempe. Next to the mosque complex are a Lebanese restaurant (named Carthage), a barber, and a bazaar with a halal butcher and other services, a combination that reflects the communal character of Islam (compare Satha-Anand 1998). Clearly the center also serves as a cross-cultural meeting place. The social capital that is invested in and arises from this complex is typically cross-cultural, a reflection of Islam being a cross-cultural religion. Thus it refers not to ethnic social capital, but to rainbow social and cultural capital. Smaller mosques in the Islamic diaspora can cater to Sunni Muslims of specific national origins (for example, mosques in Amsterdam neighborhoods for Moroccans, Turks, or Surinamese, but rarely for all). Even then they are not "ethnic" because Moroccans from different parts of Morocco may frequent the mosque (Nederveen Pieterse 1997b).

The Detroit suburb of Dearborn, Michigan, known as the Ford Motor Company headquarters, now ranks as a center of "Arab America" where 275,000 Middle Easterners have settled, the largest concentration of Arab Americans in the country. Middle Eastern immigration started with the Lebanese early in the twentieth century and has since brought immigrants from every country across the Middle East. While no more than half of these are Muslims, a new mosque complex is being built, spread over ten acres and billed as the largest mosque in the country. Services at the existing mosque "draw a diverse crowd of devotees from throughout the area, including many Americans whose ancestors emigrated from Europe or Africa long ago and who have since converted to Islam" (Lee 2000). In such cross-cultural and cross-national conglomerations as East Dearborn, ethnic economy becomes useless as an analytical category and is clearly much too narrow. A different conceptualization is needed.

It follows that we need to question and open up the notion of ethnic economy itself.[5] It is true, of course, that ethnic groups have been formidable social, cultural, and economic forces, past and present (Kotkin 1992), but ethnos, *ethnie*, or ethnicity simply mean "people," and there are people and people, or many different peoples within a people (di Leonardo 1984). A shortcoming of the ethnic economy approach is that, like most approaches that deal with ethnicity, it ignores the hybrids, the in-betweens. In reality, there are no neat boundaries between ethnic groups; the boundaries are typically fuzzy and permeable (Lowe 1991; MNghi Ha 2000; Nederveen Pieterse 1995, 2001b). Thus, many immigrant entrepreneurs who are labelled ethnic are in reality hyphenated and mixed, and on this ground alone ethnic economies tend to be hyphenated economies. Furthermore, an "ethnic economy" is not necessarily an economy with a degree of interconnectedness or integration, but rather a random set of businesses.

It may be a different case if we consider ethnically diverse societies (i.e. diverse not as a result of recent immigration). Here, ethnicity and ethnic social capital *may* be relevant terms with the proviso that there are many different varieties of ethnicity here too (Nederveen Pieterse 1997a). Besides, of course, cross-cultural relations count here as well. In his fine study of the transnational informal enterprise of Ghanaians worldwide, Amponsem rejects the term ethnic economy and opts for *embeddedness* instead.[6] He distinguishes

"ethnic" from "non-ethnic" enterprise by the degree of embeddedness
of organizational strategies in informal personal networks, trust and
social relations . . . Ghanaian immigrant business strategies and
practices are highly organized along crosscutting and cross-community
ties, social trust and informal relations.

(Amponsem 1996: 213)

What sets them apart from mainstream firms is that the latter are
organized along formal and contractual relations. "Given that the
differences are contextualised and analysed in the dualist model
rather than the interface process, the difference is conceptualised as
'ethnic' and 'non-ethnic' enterprises. The ethnic economy discourse
is therefore another dualist dichotomy of 'otherness' in strategy and
practice" (Amponsem 1996: 213).

If we consider that what matters is a *difference of degree* between
the prevalence of formal and informal, contractual and non-
contractual relations in business, the discussion is set on a different
footing. Formal and contractual enterprise also involves informal
relations and implicit understandings (i.e. it is embedded in cultural
and social practices, but embedded in different ways; compare
Schmidt 2000). In the background looms the paradigm of modernity
and Parsons' pattern variables. The point of social capital, trust,
institutional density, and related notions is to open up this framework
to examine the underlying social relations that make business tick
(Portes and Sensenbrenner 1993). Yet embeddedness is too vague a
notion and not distinctive if we consider, following Polanyi, that *all*
market relations are socially (and culturally) embedded. Would it
be more insightful if instead of ethnic economy we say *immigrant
enterprise*? This leads to several other problems: does it concern first
or second generation immigrants? There is cultural segmentation
among immigrants too; they relate to widely different economic
specializations, and immigrant enterprise comes in many varieties,
as the discussion below suggests. Instead of referring to ethnicity, we
distinguish between (mono) culturally embedded and cross-cultural
social capital.

Cross-cultural enterprise

A general consideration is that immigrant economies, in order to
function, need to build ties with other communities and cultural
groups. By labeling immigrant enterprise "ethnic" and by focusing

on its informal and grassroots character, we set it apart. The major drawback of "ethnic economy" discourse is the suggestion and assumption of ethnic boundaries. Ghanaian informal enterprise reaching across the world – Düsseldorf, London, Amsterdam, Vancouver, Bangkok, and so on – involves many non-Ghanaians, formally and informally. Informal business relations are not confined within ethnic boundaries. Amponsem describes the social relations of Ghanaian traders in Bangkok as follows:

> Successfully living in an isolated Bangkok hotel for four weeks, without family and with limited contact with the foreign social environment, is only made possible for the trader through the social interaction and the family atmosphere created together with other traders and migrants – a "little local community." It is usually a scene of sharing and interaction reminiscent of a social gathering of "communities" as Ghanaian traders from different parts of the world meet their counterparts from Nigeria, South Africa, Zaire, Mali and Guinea, for example, at the lobby of the Top High Hotel in the Pratunam area in Bangkok.
>
> (1996: 95)[7]

If Dearborn, Detroit for "Arab America" and Jackson Heights in Queens, New York for South Asians are spectacular examples of cross-cultural agglomerations, the principle of cross-cultural relations *across* immigrant groups holds much wider. It applies to groups such as the overseas Chinese, among whom immigrants from different regions often mingle (see Lin 1998; Liu 1998; Minghuan 2000). It applies to settings such as Amsterdam Southeast where Surinamese, Antilleans, Moroccans, Ghanaians, Ethiopians, and other Africans each tend to have their own circles, but also mingle (Hannerz 1992); or to Mount Pleasant in Washington, DC, Spanish Harlem in New York and East LA, and their mixed Latino presence. Labor, training, customers, supplies, credit, and possibly accountants, solicitors, and real estate necessarily bring immigrant enterprises into contact with many other networks. Neighborhood and social life are other factors. All this tends to be concealed from view if the heading is "ethnic economy," and thus these links are under-researched. Meanwhile it is also true that, when it comes to issues that represent deeper forms of integration, such as unionization and health insurance, immigrant groups often appear to be insular (e.g. on California see Milkman 2000).

Labor is a keynote in the definition of ethnic economy. "The ethnic economy is ethnic because its personnel are co-ethnics" (Light and

Karageorgis 1994: 649). This is a narrow criterion,[8] but even by this criterion ethnic economies may be a shrinking phenomenon. With growing migration in conjunction with a hierarchy among emigration countries, cross-cultural employment has long been on the increase. Thus, Japanese restaurants with Korean waiters are common, and examples along these lines are abundant. Job seekers in the culturally segmented labor market of Toronto use both ethnic and interethnic circuits which fulfill different roles. Using interethnic ties helps people gain access to diverse resources beyond their homogeneous networks. Access to social capital beyond the ethnic group's boundary principally benefits members of the ethnic group who are concentrated in lower paying jobs, while for members of mainstream, higher status ethnic groups, using intra-ethnic ties is associated with higher income. The advantage of using interethnic ties is conditional on the socioeconomic status of job seekers and job contacts: if the contact is with higher status ethnic groups, the use of interethnic ties is more rewarding than are ties with members of lower status groups (Ooka and Wellman 2006).

Los Angeles ranks as "the sweatshop capital of the United States" (Bonacich and Applebaum 2000). Here, according to a *Los Angeles Times* poll, "minority-owned firms tend to hire within their own ethnic group," but actually the patterns diverge. Businesses owned by Latinos in Los Angeles county describe their workforce as three-quarters Latino; 41 percent of black businesses report a mostly black workforce; and, of Asian firms, one-third employed mostly Asian workers and almost as many had a mostly Latino workforce (Romney 1999). The latter pattern of cross-cultural (or interethnic) employment, such as Asian garment manufacturers in California employing Mexicans, is confirmed by further research (Light, Bernard, and Kim 1999). The distinction between exploitative and non-exploitative trust is not likely to coincide with the line separating cross-cultural and same culture employment. Is trust less exploitative when employer and employee share the same national origin? That would overestimate the homogeneity of national origins: among South Asians, caste differences crosscut national and regional identities; among other migrants, class, region, and religion enter the equation.

In East San Jose, the Latino shopping center Tropicana has in recent years seen an increase of immigrant Vietnamese business owners who now own nearly one-third of the shops. In a new shopping center across the road, El Mercado, a deliberate attempt is now being made to blend Mexican and Asian cultures. Art is being used as a tool to blend the communities (for example, with an exhibition on marketplaces from Mexico, Vietnam, Nigeria, and Portugal). A Vietnamese architect comments, "We need to create a myth, the kind of myth that highlights our relationships and the good things between the communities" (Melendez 2000). Thus, cross-cultural commerce is a growing trend and an emerging theme. Part of this is due to the rise of "ethno marketing" as a function of growing multiculturalism in many countries (e.g. Halter 2000; also Radcliffe and Laurie, this volume). For commerce in cross-cultural settings, the importance of cross-cultural skills such as language is also being recognized.

Meanwhile, research on ethnicity and ethnic economies has generally concentrated on the *inward* character of ethnicity to the neglect of relations with the outside world. But how do immigrants relate to the wider economy and society? They function commercially and as entrepreneurs by acting as go-betweens or by integrating. Armenian businessmen in Europe and North America, Lebanese contractors and shopkeepers in West Africa, Chinese *tokos* (shops) in the Caribbean, Chinese businesses in Hungary and Spain, Surinamese stores in the Netherlands, Korean grocery stores in the USA, Palestinian stores in California, Indian corner dairies in New Zealand. All deal with suppliers, customers, and employees of different ethnicities, whether local or of other immigrant communities. Operating in an intercultural space affects the consciousness and identity, habitus and business practices of immigrants, as research among Colombian businessmen in the Netherlands shows (van Cotthem 1999). In Germany, Turkish businesses employing more than ten people have increased to almost 5,000 in 1998; Turkish businesses include not only retail and restaurants (61 percent), but also service, manufacturing, and construction sectors (27 percent). A Turkish enterprise that began as a travel agent for guest workers is now Germany's eighth largest travel agent, with an annual revenue of DM914 million (Richter 1999). In these cases, family or co-ethnic labor and credit may often play apart, but by no means across the board. The trend in several immigrant economies is towards greater use of bank credit.[9] A specific factor in the case of Muslims is the need for interest-free loans, for business and mortgages, which puts many in touch with a cross-cultural circuit of *hawal* bankers.[10]

Immigrant enterprise is therefore a wheel within a larger set of wheels. Cultural social capital functions, and over time can only function, as part of cross-cultural social capital. Immigrant business includes several varieties:

- Immigrant business catering to same nationality immigrants. *Not* the same ethnicity because that would be too small a market.
- Immigrant business catering to other immigrants. For example, a Dominican grocery in California selling Mexican products to Salvadoran customers.
- Immigrant business serving a niche market. For example, French hairdresser, Italian pizzeria, Chinese restaurant, Korean contractor in New York. In this case, ownership, management, labor, supply, and credit may or may not follow immigrant or national origin connections. Within this pattern, there are many variations. One variation is an immigrant business acting as intermediary between immigrants and locals, for instance in labor recruitment, contracting, ethno marketing or crime.
- Immigrant businesses from diverse origins clustered together, either by tradition or by design, as in the recent trend of "ethnic shopping malls" from Toronto to Amsterdam (Choenni 2000). (Note that "ethnic" here has the popular meaning of non-Western and refers to the combination of various cultural groups.)
- Immigrant business catering to local customers. In other words, a business that has entered the mainstream: while different in national origin, it is not necessarily different in business practices.
- A second- or third-generation immigrant business. Now the business may either continue to occupy a niche market using national origin for sign value (the Jewish deli in New York) or national origin may fade into the background. In terms of business practices and ownership (for example, a joint venture with nationals or being traded on the stock exchange), it may become indistinguishable from local enterprise. A Chinese-owned garage in Jakarta may differ from other garages only in ownership or management.

This is what a short-term typology yields. The picture changes further if we consider the *longue durée*.

In the *longue durée*

For obvious reasons, the cross-cultural dimension acquires greater depth the longer the period of time we consider. That change across generations makes a huge difference in immigrant careers is well known, yet most current research does not involve intergenerational data and leads to narrower conclusions than if we widen the perspective over longer time spans by taking into account historical research (compare Nederveen Pieterse 2000). Immigrant economies are widespread and as old as the trading diasporas and the combination of commerce and migration. The major varieties of immigrant economies distinguished in the literature are: first, minority or immigrant enterprise; second, trading or middleman minorities; and, third, enclave economies. Minority enterprise and commerce comprise the general category and a common phenomenon, as the study of the economic history of virtually any region shows. Enclave economies or immigrant businesses catering only or mainly to customers of the same ethnicity are probably relatively rare and limited; trading or middleman minorities are much more common. The Jews in Europe are the classic example. The collaboration between the Chinese and the Spanish in the Philippines, between the Chinese and the Dutch in Indonesia, between the Chettiars from Madras and the British in Burma, and between the Ismailis from Gujarat and the British in East Africa are other examples. Closer examination shows that, before the minority group was recruited for a particular function and assumed middleman status in the interstices of colonialism or empire, it was usually already present and active in the region. Regarding the Chinese in Manila, "the Spanish expedition which arrived at Manila in 1570 found four Chinese junks in the harbor. Manila, the Spaniards reported, was large and carried on an extensive trade. In the town lived forty married Chinese and twenty Japanese" (Dobbin 1996: 21). The Spanish built on the Chinese junk trade between Manila and the Fujian province and wove this regional commercial network into their own growing intercontinental galleon trade. The same pattern applies to the Parsis, who were such important brokers for the British in their commercial activities and empire building in India. Before being enlisted into collaboration by the British, they were already active as a commercial minority in the region. The presence of Parsis on India's west coast goes back to the ninth and tenth centuries when, due to Arab competition in the Persian Gulf, they moved the center of their activities eastwards:

> Thus the Parsis should not be seen as a refugee community settling down in India as agriculturalists and weavers, woken to commercial life by the European East India Companies, but rather as having much earlier developed a new trading diaspora between the Arab-dominated Middle East and Hindu India.
>
> (Dobbin 1996: 79)

Immigrant economies are embedded within intercultural economies.
For immigrant enterprise to be successful, entrepreneurs must be
at least bicultural. The Chinese diaspora in Southeast Asia and the
Pacific has been able to prosper thanks to its capacity to integrate
and build relations with the wider environment in language, cultural
skills, and the weaving of relations of reciprocity and trust. This is
confirmed by the emergence over time of mestizo groups such as
the Chinese Mestizos of the Philippines (later the Catholic Chinese
Mestizos) and the Peranakan Chinese in Java and Malaysia and
their conversion to Islam (*peranakan* literally means "child of the
country") (Wertheim 1964, 1978). This does not match Furnivall's
classic description of colonial plural society in Java and Burma
(1939), according to which different ethnic groups met only in the
marketplace where conversion and intermarriage did not take place.
The Chinese Mestizos in the Philippines are nowadays deeply
integrated and typically interact with other businessmen not on an
ethnic footing, but as members of a wider business community that
meets in settings such as the Lions and Rotary clubs.[11] Peranakan
Chinese have at times been integrated in multiple cultures at the
same time, for instance in the multicultural East Indies: Christian,
Muslim and Javanese (Taylor 1986).

The literature distinguishes between *political incorporation* of
immigrants – such as strangers in the Buganda kingdom in East
Africa who attach themselves as clients to district chiefs or to their
subjects, delivering tribute in kind or labor (Obbo 1979) – and
cultural incorporation – the adoption of language, customs, dress,
mode of livelihood, fictive kinship and religious practices, such as
strangers in Central Africa (Wilson 1979).[12] The relationship between
colonialism and trading minorities can be considered a specific type
of political and economic incorporation of immigrants. An asset
of the Chinese has been their readiness to integrate with the native
society and adopt the local language and religion (see Dobbin 1996:
64; Kwok Bun 2000). This also applies to their migration within
China.[13] Their capacity and willingness for intercultural adaptation
itself stemmed from previous experience with other trading
diasporas.[14]

> A large number of Chinese settlers were converted to Islam. Having
> come largely from Fujian, they not only found it advantageous to
> adopt the predominant religion of the Javanese port towns, but in fact
> were familiar with the role of Islam in Fujian's trade. In Quanzhou,
> Fujian's most important seaport by the late thirteenth century, both

trade and administration were dominated by foreign Muslims and an
Islamic diaspora promoted trade with the rest of Asia.

(Dobbin 1996: 47–8)

This suggests that intercultural enterprise is itself part of a *chain
of diasporas*, each imparting skills, examples, and networks of
cross-cultural intercourse. As indicated above, the classic middleman
role in colonialism is often a specification of an earlier commercial
presence and activity. Phoenicians in Carthage and Spain; Jews and
Greeks from ancient times onward; Arabs, Persians, and Parsis partly
in their footsteps; and the Chinese diasporas in their turn, along with
Indian and Malay diasporas thus form an interlinked series or chain
that stretches far back in time and widely across space. Thus, for
centuries Christian Armenians were the *trait d'union* in the silk trade
between the Safavids in Persia and the Levant (Matthee 2000). This
brief gloss leaves out many other trading diasporas and networks –
witness the history of cross-cultural long distance trade (Curtin 1984;
Stearns 2001). How deep in time some of these networks run is
suggested by the traces of trade found between ancient Egypt and
the Harappa culture of Mohenjodaro, and of trade with the Romans
found in Cochin on India's west coast.

Chain and network migration are familiar themes. In addition, a
multicultural history serves as a skill and cultural capital among
immigrant groups. In explaining the powerful influence of Middle
Eastern entrepreneurs in various industries in California, built up
over a short period of time, Kotkin (1999) notes that "particularly
Jews, Arab Christians and Armenians, have a long history of being
minorities in great polyglot cities of the Old World: Beirut, Tehran,
Jerusalem, Cairo or Damascus." Coming back to Fujian, the region,
then and now, is not among China's poorest, but has developed
an emigration culture that goes way back in time. Fujian is now
estimated to send around 100,000 emigrants abroad every year
(Deutsche Presse-Agentur 2000). The case of human smuggling
that had tragic consequences in Dover, England in the summer of
2000 (with the accidental death by asphyxiation of a group of illegal
migrants secreted in a truck) also concerned migrants from Fujian
province, as do several other episodes of illegal Chinese migration
into Europe and North America.

Ulrich Beck (2000) speaks of "place polygamy" and Pico Iyer (2000)
charts the lives of "global souls." Cross-border and transnational
social relations are growing in density and importance, and these

increasingly complex relations cannot be understood without recognizing multiple identities. For instance, the identities of settler and sojourner are not mutually exclusive (a point made by Chen [2000] in describing the trans-Pacific character of the Chinese presence in San Francisco). Akio Morita, the late Sony chairman, argued that "insiderism" is a necessity for multinational corporations: multinationals can only be successful if they become insiders in the host economies and societies, so they must "look in both directions" (Ohmae 1992). Migration history suggests that "insiderism" is common and has deep roots in time. What is now called "glocalization" (after another Japanese expression) has been common practice in the historical chains of trading diasporas.

Implications

Just as the *ethnic enclave economy* approach emphasizes clustering and boundaries in *space*, the ethnic economy approach emphasizes difference and bonding along lines of *culture*.[15] When research in this field takes into account cultural difference, it is concerned with "ethnicity." The hurdle of "ethnicity" entails an overriding preoccupation with the difference between mainstream and "other" identities to the relative neglect of crosscutting relations. In effect this involves a twofold reification: cultural difference is reified as "ethnicity" and ethnicity is reified as "ethnic economy." If we look closer, there are ample instances of intercultural economic activity and ample literature as well. Yet, by and large, this remains under-theorized and under-represented, and thus these instances do not reach the threshold of awareness in research or policy.

This review of various settings and types of immigrant economic activity, past and present, has drawn attention to their cross-cultural or multiethnic character. Reviewing the arguments presented and focusing on the key hurdle of "ethnicity," the concept "ethnic economy" involves the following problems:

1 "Ethnic" as an ethnocentric term (merely denoting distance from European or Western culture) must be distinguished from "ethnic" as an account of cultural embeddedness. However, since "ethnic" is often used loosely in many different senses (i.e. emic and etic or by outsider and insider standards), it may not be feasible to maintain such a distinction.

2 Using "co-ethnic labor" as the criterion to define ethnic economy is vague and too narrow. Customers, credit, suppliers, ownership and location are other relevant criteria.

3 If it is possible to verify whether what seems "ethnic" really is ethnic or is culturally embedded, its significance should not be taken for granted at the risk of stereotyping. Therefore, a more effective and neutral distinction is that between monocultural and cross-cultural social capital.[16] Considering that cultural and group boundaries are typically fuzzy and fluid, this distinction should not be given exaggerated weight.

4 Culturally embedded norms and social networks may indeed be significant, but immigrant economies also require cross-cultural social capital to function. Twinning social capital and cultural difference yields the following distinctions:

   • bonding social capital or close ties, which may be culturally embedded;
   • bridging social capital or loose ties at the same socioeconomic level, which may be culturally embedded and/or cross-cultural;
   • linking social capital or ties with others at a higher socioeconomic level may be culturally embedded and/or cross-cultural. Thus "ethnicity," as suggested by the ethnic economy terminology, may be relevant with regard to bonding social capital, but not necessarily with regard to bridging or linking social capital (a precis is given in Table 6.1).

Beyond "ethnic economies" are rainbow economies. This can be summed up in two points. Cultural difference does inform social capital, but ethnicity is not helpful as a terminology and analytical category. The "ethnic economy" concept must be rejected because what matters generally is not ethnicity but nationality or varieties

Table 6.1 *Social capital and cultural difference*

Types	Meanings	Cultural variations
Bonding	Strong ties among close relations	Possibly culturally embedded ('ethnic')
Bridging	Weak ties among people from diverse backgrounds, but similar socioeconomic status	Culturally embedded or cross-cultural ("ethnic" or interethnic)
Linking	'Friends in high places'	Culturally embedded or cross-cultural

of national origin. Moreover, in immigrant enterprise, social capital is not merely internal to the immigrant community, but spills over cultural boundaries. Immigrant economies are often blended or rainbow economies that rely on cross-cultural resources and social networks. Thus, for bonding social capital to deliver requires bridging social capital. A third variable (linking social capital) relates particularly to home country resources. What matters is neither the situation of full separation behind cultural boundaries ("ethnic economy" and multiculturalism as a mosaic of ghettos) nor the situation of full assimilation (cultural boundaries don't matter), but rather the in-between zone that Portes (1996) refers to as "segmented assimilation." Most research on social capital tacitly assumes or overtly focuses on cultural boundaries. It would be appropriate for research to pay as much attention to bridging social capital, in the sense of loose relations across cultural boundaries, as to bonding social capital within cultural boundaries. An implication for policy is not to rely merely on ethnic or immigrant social capital, but to take into account and enable cross-cultural relations (i.e. not simply within but between immigrant groups and between immigrants and others).

Policy implications

If the intellectual importance of social capital is to bridge diverse disciplines (sociology, political science, economics), its policy significance is to link civic cooperation (sociology), democratic governance (political science), and economic growth (economics). Thus, Putnam's study of administrative reform in Italy (1993) points to the importance of civic traditions and local democracy for administrative and economic performance. Similar implications follow from studies of "institutional densities" in geography (Amin and Thrift, 1993, 2004), of industrial clustering and districts as exercises in collective learning (dei Ottati 1994), government enablement (Helmsing 2000) and intersectoral partnerships of local government, firms, and NGOs (Brown and Ashman 1999). These lines of research involve interesting takes on social capital. The recognition of good governance as a social capital asset holds significant policy implications beyond the stipulations of the World Bank and the refrains of Washington rhetoric. It suggests that local democracy is not merely desirable on political or moral grounds, but can also be economically productive; social capital can serve as a bridge between social cooperation, progressive politics, and forward-looking economics. However, this uplifting story falters when it comes to ethnic diversity. A World Bank report notes: "Recent research has found that ethnically fragmented countries tend to have slower growth, lower levels of schooling, more assassinations, less financial depth, and higher deficits" (World Bank 1998a: 18).[17] Current research is concerned with examining "how political institutions can be reformed to secure the benefits of ethnic social capital while diffusing the costs" (World Bank 1998a: 18). It would be interesting to examine cross-cultural social capital in these settings.[18]

A point often made and a fundamental consideration for policy is that social capital without resources is a cul de sac. At the time when social capital was becoming a fad in addressing urban poverty in the USA, its downside was also becoming apparent:

> There is considerable social capital in ghetto areas, but the assets obtainable through it seldom enable participants to rise above their poverty . . . the call for higher social capital as the solution to inner city problems misdiagnoses the problem and can lead to both a waste of resources and new frustrations. It is not the lack of social capital but the lack of economic resources – beginning with decent jobs – that underlies the plight of impoverished urban groups.
>
> (Portes and Landolt 1996: 20, 21)

This raises the question of whether indeed social capital *is* capital (Robinson, Schmid and Siles 2002). At any rate, the importance of resources varies according to the circumstances. Research bears out that "the relative importance of investments in physical capital and schooling appears to vary with the extent of social development. In particular, schooling is important at low levels of social development, but physical capital becomes more important at higher levels" (Temple and Johnson 1996: 41).

This must be factored into an understanding of different immigrant economies. East Asian immigrants in North America are backed by the financial hinterland of the Tiger economies: immigrants from the Middle East can tap into oil revenues or remittances of relatives working in the oil economies. Backed by financial capital from overseas, relayed by regional banks, they can buy into prosperous markets. In free enterprise capitalism, without government support for job creation, these groups can create their own jobs by buying stores or businesses. The link between capital and migration is clearly on the map in relation to Chinese and Taiwanese immigrants (Tseng 1994, 2000). In Los Angeles alone, the home of the largest Korean population in the USA, there are seven Korean American banks (Andrejczak 1999). Start-up capital is a component that African immigrants lack. The relative poverty of much of sub-Saharan Africa does not provide them with a financial hinterland to fall back on. The same applies to many Latin Americans and South and Southeast Asians.[19] For African Americans too, there is no financial depth backing them.

A further consideration is that neither social capital nor economic resources may deliver in the absence of *capable agency* (or human capital) (Krishna 2001). Therefore, another pertinent resource is education, which is more advanced in some regions than in others. Due to their educational background, Indian immigrants in Britain and North America have been able to enter the professions early on, particularly in the fields of medicine, education, and software.[20] In other words, immigrants' differential economic and social performances are also functions of differential country resources and, of course, the immigrants' class location in the country of origin. Affirmative action and multiculturalist policies usually focus on supporting immigrant communities or on relations between immigrants and the host community.[21] The present argument suggests a further angle, namely reckoning with cross-cultural relations not only between immigrants and locals, but also among and across different immigrant groups. In urban policy, taking into account and, under some circumstances, fostering such cross-cultural relations may be considerations.

What is underway implicitly in areas such as planning "ethnic shopping malls" may become an explicit policy consideration. There are ample situations where such an approach is in fact being implemented (see d'Andrea, d'Arca, and Mezzana 1998). One example is the recent development of a local exchange trading system (LETS) of local inhabitants and asylum seekers in Woudrichem, a small Dutch town. The system involves asylum seekers providing services (haircuts, food preparation, household and garden work, party catering, drawing lessons) and Dutch locals offering goods (used bicycles, computers, videos) and services (cab rides, language and orientation lessons). Since asylum seekers may stay for long periods but do not have work permits, this system integrates them into the local economy and community without payment of money. An economist from Rwanda administers the system. A cafe night every two weeks serves to facilitate contacts. This initiative involves the generation and deployment of cross-cultural social capital, both among asylum seekers (from many different parts of the world) and between asylum seekers and locals. Policy can therefore make up for specific social capital shortfalls. Thus, under some circumstances, specially targeted start-up credit facilities could be provided to immigrant entrepreneurs who do not come from rich hinterlands (i.e. who are short of linking capital). To use a grand term, we could call this a policy of cross-cultural democratization and a step from multiculturalism to interculturalism. The forms this might take would differ on a case-by-case basis.

Notes

1   "Europe's Immigrant Entrepreneurs Are Creating Thriving Businesses – and Thousands of Jobs" (*Business Week*, February 28, 2000); "Today's Refugees Are Europe's Future Assets" (*Guardian Weekly*, April 18, 1999). "Immigrants Create Wealth; They Add Critical Ingredients to the Mix that Generates Progress. They Tend to Take Risks" (*Sunday Telegraph*, February 6, 2000); "Keeping the Hive Humming: Immigrants May Prevent the Economy from Overheating" (*Business Week*, April 24, 2000).

2   In the UK, Germany, Australia (Saunders 1999), and Israel (Hoffman 2000). On Germany, see Finn (2000); on Switzerland, see Piguet (1999).

3   For critical reflections on the genealogy and use of social capital, see Woolcock (1998), Fedderke, de Kadt, and Luiz (1999), and McNeill (1996, 2003).

4   Smart (1993) criticizes Bourdieu for applying inconsistent definitions of different kinds of capital.

5   Criticisms of the ethnic economy approach are growing as the notion suffers from culturalism and ethnicism. Noting several problems in the sociology of the ethnic economy, Cobas (1989) mentions a contradiction between the stranger hypothesis and the protected market hypothesis, while MNghi Ha (2000) draws attention to the underestimation of hybridity (compare Werbner, 2000).

6   Following Granovetter (1985), many others also refer to embeddedness in this context, such as Portes and Sensenbrenner (1993), Portes (1994,1995), Rath (2000), and Schmidt (2000). See also Chapter 1, (this volume); James (this volume).

7   Transnational informal enterprise is not necessarily cross-cultural as, for example, Portes (2000) describes.

8   Light, Bernard, and Kim (1999) adopt the term immigrant economy instead of ethnic economy if there is "non-ethnic labor," in other words labor is the key criterion.

9   A survey of Latina business in Orange County, California shows that "Few borrowed money from banks to start businesses, but the percentage of those

with bank credit has grown" (Norman 2000). Boubakri (1999) shows the same for immigrant enterprise in France. Meanwhile, other reports point out the lack of access to bank credit for rapidly growing immigrant enterprises (Aguilera 2000; compare Fisman 2000).

10 *Hawal* banking is a financial infrastructure using informal networks among Muslims worldwide. Reports on *hawal* banking circuits in the Washington, DC area are by Noguchi (1999).

11 Oral information from Peter Chua (Santa Barbara, California).

12 Shack (1979) discusses varieties of incorporation of strangers in sub-Saharan Africa.

13 "They [the south Fujianese] were prepared to merge with the social and economic networks of the host city and in Shanghai, for example, many Fujianese were regarded as having 'become local people,' enabling them to penetrate local networks in most business circles" (Dobbin 1996: 64–5).

14 Compare with Tambiah's (2000) fascinating account of the intergenerational Man lineage of the Chinese.

15 The ethnic enclave economy is a predecessor of the ethnic economy approach; compare Light, Sabagh, Bozorgmehr, and Der-Martirosian (1993) and Werbner (2001).

16 I deliberately refrain from defining "culture" in this discussion (See Chapter 1, this volume; Watts, this volume). At a subtler level of analysis, the distinction between mono- and cross-cultural enterprise (and social capital) is up for discussion. Cultural differences run, of course, also among nationals, for example by region. Ethno marketing and import and export trade are also cross-cultural – witness the literature on "international manners" in business and on varieties of capitalism.

17 Note the charged choice of words of ethnically fragmented rather than ethnically diverse societies.

18 In my own work, I ask "Under what conditions do we get widening circles of social capital?" (Nederveen Pieterse 2001a: Chapter 8). I argue that government can play a facilitating role in the form of managed pluralism (compare Midgley 1995; Gold and Light 2000).

19 This may be an explanation for recent reports that new immigrants in Canada and the USA are not doing as well economically as previous waves; they may have less access to financial resources in the country of origin. It might also explain why Indonesian immigrants in the USA lag behind others.

20 In Britain, South Asians "[a]lthough they represent just under 3 per cent of the population, . . . provide about 16 per cent of the total number of GPs, nearly 20 per cent of hospital doctors, and about 12 per cent of pharmacists. They own just over 50 per cent of the 'cash and carry' shops and just over 55 per cent of the independent retail trade" (Parekh 1997: 65).

21 The informal economy is generally not a high priority policy area (see Rath 1999).

7 Social capital as culture? Promoting cooperative action in Ghana

Gina Porter and Fergus Lyon

In this chapter we focus on culture and its complex interconnections with the concept of social capital. Our study is set in the context of recent preoccupations of the World Bank (followed by other donors) with the concept of social capital itself and the related construct that it can be built in order to promote economic growth and development. The adoption of the social capital concept is perhaps the closest that the World Bank has come in recent years in its recognition of the potential linkages between local cultures (notably cultures of enterprise), economic growth, and development. That is not to say that we agree that culture *is* social capital – or vice versa. Far from it, indeed! Our thesis is rather that the World Bank has taken up social capital in a highly essentialized form – as group cooperation per se – in its development initiatives, while at the same time congratulating itself on its adoption of a more culturally (as opposed to economistically) oriented development paradigm. The Bank's conflation of social capital construction with group activities, and its consequent efforts to promote development through supporting group-based initiatives, far from illustrating a cultural turn in its development thinking, arguably reflect a continuing lack of sensitivity to cultural diversity and the specific geographical contexts within which diverse cultural registers (elite, popular, and youth cultures, among others) evolve and interact (see also Nederveen Pieterse, this volume).

Following a brief introduction to the links between concepts of social capital and culture, we review recent development problems and donor activities in Ghana. We then present two rural case studies of group-based development interventions in Ghana's coastal savanna to illustrate our argument that while culture is complex, multifaceted, and inextricably linked with place and time dynamics, recent development interventions seemingly emanating out of donors' desire to build social capital have been based on very poor conceptualizations of culture.

Social capital, culture, and development

Social capital is, as Harriss (2002) observes, a slippery concept. Widely publicized through the work of Putnam (1993, 1995), it focuses on the potential benefits of associational life and collective action. The positive value of social relations built on trust, norms, and networks is a central theme, though precisely how these relations are initiated and sustained, and by whom, lacks explication. Nonetheless, social capital is assumed by much of the donor community to bring voluntary cooperation that leads to improved welfare and economic performance (Barr and Toye 2000). This focus on social interaction as a positive win–win situation of empowerment and inclusion (DeFilippis 2002) has been extremely seductive to development specialists mired in their persistent failure to find solutions to seemingly intractable poverty problems. There has been a growing view that interventions are possible which can harness underlying social forces and energy in society and thus expand social capital and correct for state and market failures (Mayer and Rankin 2002). Consequently, social capital "has become one of the central organizing themes in global development work" (DeFilippis 2002), in no small part due to its adoption by economists at the World Bank, where it was identified in a 1997 publication as a developmental "missing link" (Harriss 2002: 9, 82).

The linkages between social capital and culture are complex. Much depends, of course, on how we define culture, another notoriously slippery concept (Mitchell, D. 2000). Here, drawing on Rankin's (2003) efforts to combine standard recent anthropological interpretations of culture (as opposed to cultural studies perspectives) with geographical concerns for the significance of place and scale, we conceptualize culture as a grounded dynamic and symbolic entity. It encompasses people's actions, shared beliefs, values, behavior, and the meanings they assign to their world and those actions.

This includes recognition of the potential of both history and multiple, intersecting spatial scales, to shape (and be shaped) by those actions and meanings (for further discussion of these themes see Chapter 1, this volume).

The concept of social capital has opened an opportunity for economists and non-social scientists to consider the concept of culture. However, the ways in which the terms social capital and culture are interpreted requires examination. Putnam defines social capital as trust, networks, and norms each of which is underpinned and shaped by cultural contexts (Putnam 1993). Putnam examines the spatial differences of social capital by focusing on accumulation of social capital within locally based national-ethnic populations (compare Watts, this volume; Nederveen Pieterse, this volume). From this perspective, the concept appears an attractive route to benefit poor people: thus, logically attractive to many development organizations and their staff.

However, recent analyses of the social capital model have raised numerous questions. These center around debates as to whether social capital is a way of opening economistic approaches to development to issues of cultural specificity, or whether it is a way for economics to colonize other social sciences. Fine's argument that the model has a tendency to ignore the existence of power relations and conflict, that "social capital is ahistorical and asocial, so it is complicit with mainstream economics" (Fine 2002: 796–9), has gained particularly widespread support from academic researchers. Meanwhile, a number of empirical studies have illustrated the "down-side" to social capital in poor countries: the exclusions, clientalism, and backlash that have accompanied many supposedly pro-poor grassroots development projects (for example Mayoux 2001; Momsen 2001). These studies graphically remind us how effectively power relations can distort the "harmonious sociability" romantically ascribed to communities, severely reducing the potential to build extensive networks of trust (on expectations around community, see Watts, this volume). In response, proponents of social capital make the distinction between bonding social capital (within a community), bridging social capital (outside a community), and linking social capital (ties between people of different wealth and status) (Woolcock 2000; also Nederveen Pieterse, this volume). However, in practice, development interventions tend to be preoccupied with bonding social capital.

This chapter extends and refines these broad criticisms that local power relations may severely impede externally generated interventions aimed at building local social capital, by emphasizing the specific role that space and place play in shaping local cultures of cooperation or non-cooperation. We show how in Ghana, as in so many other countries' recent development programs, there has been a widespread tendency to reduce the concept of social capital to mean merely "membership in local, voluntary associations" (Harriss 2002), and then to apply this blueprint of group cooperation in an asocial and aspatial manner that ignores local specificities of place, space, and cultural context. The remarkable lack of geographical imagination evident in recent developmental applications of social capital adds yet another nail to its coffin.

Africa's "beacon of hope": donor perceptions of late twentieth and early twenty-first century Ghana

In recent years Ghana, a small West African country of around 18 million people, has arguably become nothing less than a donor's darling. A recent UK DfID document refers to it as "a beacon of hope and stability amidst a turbulent region" (DfID 2002: 7). Although continued dependence on primary products (gold, cocoa, timber) has limited Ghana's potential for rapid economic growth, it is seen to be "favored by a healthy democracy and a strong independent media" (DfID 2002: 7). Ethnic and religious tensions occasionally emerge, but strife tends to be spatially circumscribed. The successful transition from government by Rawlings' NDC to Kufuor's NPP government in December 2000 was marked as a particularly notable success.

Ghana has recorded some significant firsts in its recent history: notably, first African country to achieve independence (1957), and first African country to implement a (painful) World Bank/IMF structural adjustment program (1983). The latter program marked the start of a long relationship with the World Bank, which continues through to the present. Unlike its more truculent near-neighbor Nigeria, the Ghanaian government has mostly tried to work with donor agendas to reduce the poverty that persists across the country, but is particularly widespread in its three northern regions.

So far as expatriate staff are concerned, Ghana (by comparison with other African countries) is a very good place to be located: above all, violence is low, Ghanaians are extremely welcoming to strangers, and many donors perceive a national ethos around avoiding tension and strife (to the extent that one donor interviewed suggested that Ghanaian culture may be "dangerously confluent") (Authors' interviews with donor staff, 2003–4). The relative ease of working in Ghana with Ghanaians cannot be ignored as a factor which encourages continued donor support to the country. This perception of cultural harmony may have an important bearing on recent donor interventions to support the country's "latent entrepreneurship" (DfID 2002: 7), since the tendency to see Ghanaians as essentially collaborative also makes them supremely suited subjects for experiments in social capital building. The fact that Ghana is widely recognized and represented as a country with a rich traditional associational life (Atingdui, Anheier, Sokolowski, and Laryea 1998), gives an additional impetus to donor expectations.

Donors and development in Ghana: the role of the Local Development Groups

Local Development Groups, or LDGs, are at the heart of many recent donor initiatives aimed at building social capital worldwide, since membership of groups is commonly argued to be crucial to building trust and social cohesion (Weinberger and Jutting 2001: 1395). A review of World Bank literature suggests that group membership is directly equated with social capital (also Narayan and Pritchett 1997; Grootaert and Narayan 2001). For instance:

> To measure the density and importance of social connections in rural Tanzania in 1995, researchers asked households to list the groups they belonged to . . . They then constructed an index of social capital incorporating various aspects of membership . . . Villages rich in social capital had higher incomes than those with little. They were also much more likely to use fertilizer, agrochemical inputs, and improved seeds.
>
> (World Bank 1998b: 121)

On the ground, the potential for groups to allow rapid disbursement of funds appears to be a particular factor in their favor among donors and local NGOs alike: quotas for group formation and a "scramble for groups" have been observed in many countries (e.g. Joshi 1998 on Nepal; Quirk 2003: 156–61 on India and Mishra, Vasavada, and Bates in press). Mayer and Rankin (2002: 805) state that micro-finance models advocated by mainstream donors like the World Bank "respond more to lenders' concerns with financial sustainability than to traditions of fostering radical collective action . . . solidarity groups function foremost to cut costs and introduce financial discipline through peer pressure." Much the same comment can be made regarding donor support of LDGs.

In Ghana, the LDG has been promoted with remarkable persistence by donors, led by the World Bank. The broad donor perceptions concerning cultural harmony in Ghana that we described above may well have been an important contributory factor supporting this promotion. Community groups had been at the center of development initiatives among donors and NGOs in Ghana well before social capital was taken up as a key theme by the World Bank, but the focus on LDGs has gathered pace over the last decade, so that most NGOs in Ghana now focus the majority of their activities on groups. Donor and NGO rhetoric emphasizes LDGs as central to liberal democracy and democratic development in Ghana, but the sad reality is that numbers of groups and group members and associated loan recovery rates have become more prominent NGO targets than real improvement in living conditions or participation. Experience in the micro-credit sector in Asia seems to have a particularly strong impact on donor ideology about groups in Ghana, as elsewhere in sub-Saharan Africa. The Grameen bank in Bangladesh, in particular, is still commonly held up as a successful model, although the application of such Asian models in Ghana has been attributed by local academics to a lack of confidence in local ideas and identified as a factor which actually undermines local potential for change (Porter 2003). Senior figures in major local NGOs talk about being "rolled over" by donors because they lack the formal evidence (i.e. support from academic writings in international publications) to support their local perspectives.

Ghana might well have been expected to present a fertile seedbed for donor group-based initiatives because of its rich tradition of associational life and nationally promoted values of cooperation. However, important spatial distinctions can be drawn, for instance between northern and central Ghana and the coastal zone. These distinctions, which we suggest have been inadequately recognized by donors (and consequently their client NGOs), help shape the potential for success of group-based interventions on the ground in each region. A brief description of each zone and its history of group action helps set the national scene in which the more detailed discussion of the case study LDGs can then be placed.

In the 1980s and 1990s northern Ghana became the site of the majority of international and local NGO activity, because of its chronic poverty and limited infrastructural development. (In the colonial era, this region was viewed principally as a labor reserve for southern gold mines and cocoa farms.) Initially assistance was focused on basic service delivery, though following the mid-1990s' ethnic disturbances many NGOs expanded their focus to citizenship and ethnicity (Mohan 2002). Most NGOs working in northern Ghana now focus a majority of their activities on groups. There are many such associations: groups for agro-processing, revolving credit schemes with goats, groups for money literacy and income generation, and so on. Many of these programs are focused on women and some involve membership groups as large as 150 people.

Central Ghana exhibits rather different local cultural, political, and economic contexts from both northern Ghana and the coastal savannas. Much of the central area – notably the Ashanti and Brong Ahafo Regions which form the Akan heartland – is richer than the north: it includes major maize and export cocoa producing zones, gold mining and timber production, and has more developed communications and infrastructure. This is a region capable of substantial community mobilization without external support, related to its history of opposition to central government and the tradition of demonstrating group support through funeral attendance (Dennis and Peprah 1995). NGO activity has been very limited in the region (Kyei 1999) because donor support for interventions is focused on Ghana's poorer districts.

Further contrast is offered by the coastal savanna region where there has also been little NGO involvement, despite the fact that there are considerable pockets of poverty. There is much dependence here on semi-substance farming and artisanal fishing and some areas are arguably as poor as northern Ghana. In recent years a number of donor-led initiatives have extended into this zone (which is facing serious environmental problems) and a few international NGOs – which had previously concentrated on northern Ghana – have begun to operate more extensively here. Additionally, quite a few local NGOs have come into existence, though by comparison with northern Ghana their numbers remain comparatively low. Again, group-based activities have commonly been central to the development initiatives set in motion.

Our discussion so far has shown how donors (and consequently both local and international NGOs – who mostly depend on donors for their survival) have adopted a relatively uniform concept of cooperation in groups. We now aim to demonstrate how and why attempts to implement this aspatial blueprint have been rejected or adapted, by reviewing the diverse experiences of group formation in one region, the coastal savanna zone.

LDGs in coastal Ghana: patterns of adoption, adaptation, and rejection

Case 1: Struggling to form groups: Ghana's Village Infrastructure Project and Intermediate Means of Transport

This case study explores support for and resistance to group formation in a rural transport context. The broad donor development context is provided by the nationwide Village Infrastructure Project (hereafter VIP), which was established in the second half of the 1990s. Associated with the VIP was an Intermediate Means of Transport (hereafter IMT) project involving the Ghana Ministry of Agriculture's Village Infrastructure Project Coordinating Unit. The experiences of five villages in the Central Region not previously targeted under the broader VIP program were followed closely during the process of adoption of the Intermediate Transport project (IMT) to assess their impact within the villages. The latter evaluation was carried out as part of an action research project involving one of this chapter's authors (Porter 2002a, 2002b).

The Village Infrastructure Project (VIP) is a World Bank funded program focused explicitly on rural community groups and associations in pilot locations across the country. Project components include rural water infrastructure, rural transport (including village tracks and trails and intermediate means of transport such as bicycles and trailers, power tillers, and carts), post-harvest infrastructure (e.g. drying facilities and so on), and institutional strengthening focused on the district assemblies. While previous programs relied more on existing groups, new groups can be eligible for assistance in this program, providing they register formally and have "satisfactorily completed appropriate training in group dynamics and management through a partnership NGO." A further requirement is evidence of group savings at a "commercial bank."

We examine here those cases of local groups buying intermediate means of transport such as bicycles and trailers, power tillers, and carts through the VIP and a similar set of equipment purchased through the IMT action research project. From the outset, problems have been identified around "group formation, dynamism and cohesion" in the VIP (Anchirinah and Yoder 2000), particularly in cases where groups have been established for the purpose of receiving funds, and in those groups that share equipment and maintenance. Successful groups were identified in the VIP where farmers had established the groups themselves, such as *nnoboa* (joint farm labor) work groups. However, shifting the emphasis of some of these groups and, in particular, building new groups has presented challenges.

Interviews with inhabitants of the five villages in the action research project found negative attitudes to groups, raising questions about the reasons for such attitudes. Reluctance to form groups was found particularly in relation to perceptions about potential quarrels:

> I would not consider group ownership because . . . it always includes quarrelling.
>
> (Young woman, Assin-Aworabo)

> You would not get it [the IMT] whenever you need it.
>
> (Woman, Gomoa-Sampa)

Husbands tended to support this view, not wishing to see themselves drawn into village disputes: "if there is trouble arising she would come to me" (Young husband, Aworabo). Indeed, in the subsequent implementation component of the action research when transport equipment was offered to villagers on credit, we found that only five pieces of IMT out of 70 were purchased by groups. One of these cases was in a small village (Gomoa-Abora) where most residents are interrelated and two groups purchased and managed equipment together.

Moreover, the five groups have proved no faster than individuals in paying for their equipment. When we held review meetings with villagers in the five villages after a 16-month monitoring period, four out of the five village reviews still came out strongly in favor of individual as opposed to group activities. There was a common view among both women and men that group ownership of equipment would only be feasible if members were drawn from the same household, as otherwise there would be too many disputes about the use of equipment.

Despite this apparent resistance to groups in the Central Region, district assemblies and administrative officers in these districts have favored group loans, specifically because they argue that the group will apply pressure on defaulting individuals. Moreover, local officials argue that loans to individuals are difficult because of the need for collateral and guarantors. They are particularly positive about the reliability of women's groups in meeting repayments because of the application of sanctions. VIP staff, meanwhile, argue that on a large project like VIP, groups are necessary not just because of the size of the project but in particular because VIP are focusing on the "poorest of the poor" and individuals could not afford IMTs (Porter's meeting with VIP staff, Accra, 2000). At the same meeting, however, a Ministry of Agriculture representative from another department suggested that in the agro-processing field, individuals operated and managed the equipment better than groups, "although project staff find groups easier to manage." By August 2000, one of the two District Chief Executives from the study districts was expressing some disillusionment with group work in the VIP. He found the VIP regulations around the use of long-standing groups operating their own account "too difficult" and suggested that the project would work better through individuals.

Government and NGO attachment to groups was still strongly in evidence, despite the clearly expressed negative attitude of villagers to group formation for development projects, as illustrated in the following comments:

> Formation of strong groups is needed.
>> (Staff member, Feeder Roads Department)

> Women's groups can guarantee credit . . . already existing women's groups can easily be contacted for the use of IMTs . . . women's groups can guarantee for credit facilities.
>> (NGO regional project officer)

> Women's groups can influence others to use IMTs.
>> (Government officer)

> Women's groups are a force to be reckoned with. They can easily move to NGOs for funding. Repayment is guaranteed.
>> (NGO staff member)

One small workshop discussion continued to list the following advantages of groups (presented by a government staff member):

As a group, can influence one another to adopt the IMT; can easily
organize their members for training and education; can guarantee for
credit and ensure repayment; as a unified front can easily approach
local and international NGOs for support.

At local government level, however, some bemusement was
expressed that groups were patently *not* working in Gomoa district.
One district officer spoke out at the workshop in evident frustration,
"Why can't it work here with groups?" He received no response.

Case 2: Struggling to sustain groups: the case of inventory credit and processing groups

Farmer groups also lay at the heart of two agricultural projects
managed by an international NGO in the Central Region of Ghana.
The inventory credit program allowed groups of farmers and crop
processors to receive micro-credit from local banks using their
produce as collateral. Under this system, groups stored their produce
at an inventory site and then received 70 percent of the market price
at that time. When the market price rose after the harvesting period,
the produce could be sold and farmers were paid the value of their
crop minus the initial credit and the cost of storage. The organization
of groups was carefully supervised and coordinated by NGO staff
who played a key role in their functioning. Of crucial importance
was the fact that there were two locks for the store. The key to one
lock was held by the group, and the other key was held by NGO
staff. The banks did not appear to be willing to lend inventory credit
unless a trusted NGO was actively involved. This raised questions
regarding the sustainability of the group after the departure of the
highly trained, motivated, and mobile staff currently involved.

Interestingly, the pilot scheme to introduce this inventory credit
program and associated group marketing to (new and established)
farmer groups in selected areas of Ghana (Volta, Eastern, and Central
Regions) seems to have had least success in the coastal zones of
Central Region, though it was subsequently adopted with apparent
success further north in Brong Ahafo and northern Ghana. The
groups' success in Brong Ahafo was attributed to the "better
operation of farmer groups there than in Central Region" by the
NGO which organized this program (Interview, Accra, 1999).[1]
Some groups were able to repay the start-up loans sooner than
others, with success attributed by the NGO workers to the

behavior and commitment of the equipment operators and managers who collect fees for using the equipment and prepare the accounts.

This research demonstrates the importance of understanding how social capital is culturally specific and how the issues of trust and sanctions need to be understood as part of the cultural context. In the context of this project, the ability to sustain trust was found to be based on norms of reciprocity and the ability to exert sanctions. Norms also shaped what sanctions are deemed acceptable, and the implicit acceptance of particular types of authority such as the chieftaincy system. One of the groups had particular success, reportedly due to the specific commitment of a woman leader and to the fact that the machine operator belonged to the same church as some of the leading members (which may give additional moral pressure in his case). Groups with effective, committed leaders, and some previous experience of working together are more likely to be successful in this context. However, of the four sites established, only one was operating fully in 2004 due to the inability to invest in maintenance. In contrast, there are other types of groups in the locality that *do* sustain themselves over decades such as trader associations.

The failure of groups in the farmers' groups is attributable to a lack of conceptualization of what makes groups work and how they are underpinned by culture. In particular, the expected form of leadership and organization were transplanted from Western cooperative models (with a chair, secretary, and treasurer), while ignoring the local forms of leadership modeled on the chieftaincy system, which have been adapted by women trader associations that have a "queen" and elders. Cooperation within groups is also dependent on issues of trust. Trust is essential for cooperation because it provides the foundation on which other individuals' actions within a group initiative are accepted as being either for the overall benefit of the group or without disadvantage for the group. Trust comes about through an expectation that those being trusted will cooperate (Lyon 2000). Trust seems to grow on the back of accepted cultural norms that specific people will cooperate, keep verbal agreements and act reciprocally. In the Akan group of languages in Ghana the nearest equivalent term to trust is *gyedi* which also means confidence, knowledge of a person's ability, belief and faith, all of which in Ghana are firmly intertwined with concepts of reputation.

Trust cannot be fully understood without reference to processes of power and control (compare Watson, this volume). Power over others is utilized by individuals within the group, or by the group as a whole, to maintain control over group trajectories, and punish deviant behavior. This latter type of power (a) enables specific values of the dominant interests to become recognized as group norms (implying moral obligation and routine/habitual compliance), (b) allows surveillance of members' activities and behavior (which may ultimately either support or destroy trust), and (c) realizes threats or actual action when deviant behavior occurs. The sanctions which are ultimately imposed may take various forms, from group displeasure with the individual (expressed verbally or by actions, including apparent loss of trust and application of corrective pressure), through expulsion from the group to possible initiation of broader actions. Such actions may involve the wider community in implementing social and/or financial punishments, bringing loss of prestige and shame, a "bad name," financial loss, and even physical attack. When major sins are committed it may be necessary for the "sinner" to leave the community altogether.

Cultural insensitivity and the (mis)understanding of trust and norms

The projects described above illustrate a number of factors that shape cooperation including the key roles of risk, trust, and sanctions. We argue that these issues are not considered in the donors' use of the term social capital due to their lack of understanding of cultural heterogeneity in relation to local norms, attitudes, and history. In order to further understand these complex patterns of resistance and compliance to donor pressures to form groups in Ghana we need to consider a range of possible influencing factors, including specific agro-ecological conditions and their impact on prevailing livelihood opportunities and practices, related migration histories, the likely impact of remoteness, transport accessibility and proximity to urban areas, and varying histories of NGO intervention. All these issues shape the cultural factors which may predispose people to favor or disfavor group action.

Cooperation is sustained if there is trust or an expectation that other members being trusted will cooperate when it might be in their own interest not to. It operates when there is confidence in other agents, despite uncertainty, risk, and the possibility that those agents may act opportunistically (Misztal 1996: 18, Gambetta 1988: 218–19).

The more people work and associate together, the more trust may grow (Granovetter 1994: 463; compare James, this volume) and the more it can be used to support new types of interaction. Trust grows on the back of accepted cultural norms that people will cooperate, keep the verbal agreements, and act reciprocally.

Decisions to cooperate in LDGs are shaped by both conscious calculation and habitual action, sometimes by unquestioning compliance or obedience. Cleaver points out in a slightly different but related context, "non-participation and non-compliance may be both a 'rational' strategy *and* an unconscious practice embedded in routine, social norms and the acceptance of the status quo" (2001: 51). Sanctions are also an important part of trust relations and based on power relations, accepted values of hierarchy, conceptions of authority and economic power as discussed above. Norms shape what behavior is considered acceptable and what sanctions can be taken. These vary depending on whether the relationship is with kin, community members, or non-community members.

The issue of social capital and sanctions in kin groups raises specific questions about the ability to work with family groups. An environmental NGO working in southern Ghana has begun to move from a focus on community lands to family lands in project development, recognizing the diversity of interest within "communities." This has led to a change from supplying community nurseries to supplying family holdings with trees. However, family group activities may fail to operate precisely because it is so difficult to impose harsh sanctions on deviant members or because they fail to address gender inequalities within the family (Molyneux 2002). Kin-based groups also, by definition, bar non-family members.

However, donors fail to see how their interventions are being reinterpreted and how groups actually function in practice. Local gender relations, for instance may have a critical influence on group potential. Most development groups tend to have a (mostly) unpaid executive – leader, secretary, treasurer – which plays an important supervisory role. In Ghana, as elsewhere in low-income countries, most of the executive tends to comprise literate males. Women may be excluded explicitly on the basis that they lack writing and accounting skills, a factor which then often prevents women gaining equal benefits from group activities and may ultimately disadvantage group progress (Lyon 2003) unless individual strong and committed women emerge as leaders.

The local development group is commonly expected by donors to exist for a substantial period as a stable entity, through the (frequently slow) period of project planning and eventual inauguration to actual activities on the ground, to maintain its membership and its external alliances. Although groups are best when longlasting, benefits may simply take too long to emerge, when measured against other livelihood options, and may have been misrepresented by NGOs and District staff or misconstrued by groups. Moreover, in Ghana's traditional non-formal groups there is often a tendency for non-kin based associations to shift membership, focus, rules, and external alliances fairly rapidly in response to changing social, economic, and political conditions in a way which is not anticipated by donors in LDGs. Similar dynamism and fluidity among non-formal groups has been observed in Tanzania (de Weerdt 2001).

As local participants – district authorities, implementing NGOs and group members themselves – may recognize, the crucial factor ultimately is likely to be the power of surveillance and censure, but this component may be the one most difficult to graft artificially. Indeed, particular problems in a development context seem to stem from donor's inadequate conceptualization of sanctions in theorizing around social capital and the potential of group action. As we have illustrated in our case study, grassroots development workers are often far more aware than donors that group enterprises involve relationships which may incur social costs both for members and for non-members: peer pressure, loss of trust, and exclusion of the poorest and most vulnerable. It is not necessarily a win–win scenario and while direct and indirect costs (time, money, materials, argument, etc.) are likely to be incurred at an early stage, direct and indirect benefits (income, information, facilities, etc.) may take some years to become visible (Weinberger and Jutting 2001). The ability to impose sanctions is likely to vary spatially, with urban proximity causing particular potential difficulties (an observation supported by Freidberg [2001] with reference to problems of collaboration around Bobo-Dioulasso, Burkina Faso), a point we discuss further below.

Geography, culture, and development cooperation in Ghana

Our findings in Ghana suggest that the ability to build trust and impose sanctions varies substantially across regions, within regions, and between different cultural registers, especially where externally initiated LDGs are concerned.

The widespread existence of active donor-supported LDGs in northern Ghana appears logical in the context of that region's relative remoteness, its associated lack of livelihood opportunities, and its relatively long history (by Ghanaian standards) of NGO intervention. In this context, when an NGO arrives with the promise of largesse, and this is predicated on group formation, the pressure to cooperate will commonly be intense. If some livelihoods improve even marginally as a result – particularly those of more powerful community members – the group is likely to be sustained, or at least a semblance of group action maintained whenever the potential for external assistance appears on the horizon. Given the shortage of other local options, benefits from the group become part of an overall livelihood strategy and there can be strong peer pressure on individuals to cooperate. Additionally, in northern Ghana there may be social (as opposed to economic) benefits for women from group formation since cultural constraints can limit their activities and travel to distant places. Our research findings are supported by studies elsewhere showing that group operations may allow women a good excuse to meet and even occasional opportunities for travel to NGO offices at district and regional headquarters (Townsend, Arrevillaga, Bain, and Cancino 1995).

In the Brong Ahafo and Ashanti regions, livelihood opportunities are far greater than in the northern regions and thus the imperative to form groups from this perspective seems lower. However, there is much evidence in this region that groups have been a traditional component of the rural livelihoods repertoire. Relationships are often built up carefully over time, especially in remoter rural areas, so that community mobilization, including associated group formation, is feasible. *Nnoboa* groups, for instance, are formed by groups of farmers to undertake working parties at labor bottleneck periods, and many other groups may emerge from time to time. Thus, although there has been much less NGO intervention in this region, because of its perceived comparative wealth, traditional group formation is common (Lyon and Porter 2005). Consequently, when NGOs *do* intervene, we might hypothesize that grassroots interest in LDGs is likely to be sustained, especially in remoter rural areas. In the case of inventory credit (see case 2 p.160), this hypothesis was confirmed.

The coastal savanna zone, where much of our research was conducted, provides other lessons. In more accessible areas, and particularly those areas close to major cities and in settlements along major highways, populations are fluid, people come and go with the farming and fishing seasons and according to other non-seasonal opportunities which become available. "Straddling" the urban-rural divide has intensified as a feature of life in this part of Ghana since early colonial settlement, and has accelerated since the introduction of Structural Adjustment Policies by the World Bank in 1983. Group formation is likely to be more difficult in these locations within the coastal belt because individuals can far more easily disappear into a distant suburb or to another town where fellow group members and creditors cannot find them without considerable effort. In addition to this broad pattern of rural-urban mobility, there are substantial migrations of fishermen, their wives, and families as the fishing fleets move up and down the West African coast – from Abidjan to Lagos and beyond. Not only may these migrations further destabilize the potential for developing trust and sanctions. Such national and international migrations also make it particularly difficult for the kind of spatially based groups favored by donors to become established and develop the necessary link to local NGOs in order to obtain funding for specific activities. Particular examples of group failure were given by villagers concerning *susu* (non-formal rotational group saving and lending schemes). In coastal fishing villages like Dogo in Gomoa district and Ada Foah in Dangbe East district, the poor experience of development groups has turned many residents against group schemes in general (Village interviews 1997). It has possibly been a major cause of lack of trust in new (especially externally initiated) group enterprises among inhabitants of many other settlements in coastal Ghana.

Despite these problems, even in the coastal zone, groups may work in particular circumstances. In case study 1, family bonds were sufficiently strong in one village to overcome the potentially disruptive influences of urban proximity and allow some group purchase of intermediate means of transport. The four other IMT study settlements were located inland, off the paved road (between 8 and 25 km distant), on bad roads with relatively poor access to transport. They thus might have been expected to provide contexts relatively favorable to cooperation. However, the widespread reluctance to undertake NGO-promoted group activities among the Fanti even in less accessible areas, may lie in regional historical experience of defaulting from groups described above. In some cases communities may not have suffered actual bad experiences, but the mere publicity that cases of absconding receive often seems sufficient to deter moves towards group activity, despite pressures for group formation from NGO staff. This reluctance may be encouraged by the greater self-reliance and choice that people have due to the history of cash cropping in this region, as well as the lack of sanctions available to stop them.

Conclusion

Donor interventions to support social capital in many cases appear to be misguided. These interventions, founded upon the World Bank's limited conceptualization of social capital, emphasize a "win–win" scenario of trust and empowerment in groups, and have been essentialized as support for Local Development Groups. Many such projects end up focusing on specific groups that are easy to work with, essentially turning a blind eye to class, ethnic, gender, and other such social and political divisions in communities and reinforcing local structures of inequality and discrimination (Brohman 1996; Atak 1999; also Watts, this volume). There is growing evidence to suggest that this has been the case in northern Ghana (Mohan 2002).

Our chapter has charted how, while donors and others have been pursuing the LDG as a development tool in Ghana, the geography of cultural development has inserted a powerful, yet unanticipated by policy-makers, complication into this process. We have shown how group activities in Ghana are firmly embedded in local–regional cultural conceptions and attitudes: these may bring intra-group solidarity or they may encourage contestation and disharmony. There may thus be fundamental disjunctions between what donors expect of groups and what groups might be realistically expected to achieve (in specific cultural and economic contexts) and between donor and grassroots perspectives on how groups will operate on the ground. Given the limited geographical imagination of donor discourses concerning LDGs,[2] their failure in certain locational contexts in Ghana hardly surprises. It can be argued that this is directly attributable to limitations of the social capital literature which has not only failed to give adequate consideration to the importance of history (a now widely accepted criticism) but also neglects the crucial significance of geography (see Chapter 1, this volume).

Moreover, our chapter illustrates how the potential to build social capital is highly dependent on locational context: namely on regional cultural history, prevailing livelihoods and opportunities, remoteness, migration patterns, and a range of other (sometimes highly localized) economic or sociocultural factors (such as female seclusion and gender relations), what we might term the variations in cultural economy. These factors interact to produce specific local conditions that have the potential to either support or destabilize social relations in simple or complex ways, predisposing people to accept or reject group projects.

Building trust (and so-called social capital) commonly hinges on developing an intimate knowledge of people's character, personal and family circumstances, being able to monitor their actions and test them gradually over time, and, most critically, knowing where to find them if trust is misplaced and sanctions have to be imposed. It is not surprising then that we find cooperation between non-family groups and group action in response to external development interventions tends to be stronger in those parts of Ghana where local options are limited by local agro-ecological conditions and conditions favor close and regular interactions. All of the factors listed above have an inherently spatial component which may either favor or disfavor group action.

Returning to our coastal case study region, the net development result of failures at group formation and associated problems with disbursement of funds in coastal Ghana is probably to further encourage concentration of development funds in northern Ghana, where there is greater willingness to work in groups according to donor requirements, despite the fact that poverty in some coastal districts of Central Region is on a similar scale to that in the north. Meanwhile, in parts of coastal Ghana (and probably elsewhere) the requirement for pre-formed groups and working through groups may actually be diverting development efforts by local NGOs away from assistance to those individuals and traditional non-formal groups most needful of support, and creating further disillusionment among those communities which have been subjected to development. Such long-term outcomes illustrate the power of development concepts such as social capital, as well as highlighting the ongoing processes that underpin the reproduction of uneven landscapes of development.

Far from illustrating a genuine cultural turn in development thinking, donor adoption of social capital concepts arguably reflects a continuing lack of sensitivity to cultural diversity and the specific geographical contexts within which diverse cultural registers (elite, popular, and youth cultures, among others) evolve and interact. Economists continue to set and rule the agenda, despite the growing numbers of anthropologists and other social scientists employed by the World Bank and other donors. The consequences are well illustrated by recent donor development initiatives in Ghana. Although donors less commonly voice concerns about project replicability and generic approaches than they did in the late 1990s, there nevertheless appears to be a persistent reluctance to take on board cultural diversity and its developmental implications.

Notes

1   The same international NGO had also been supporting mechanized oil palm processing projects in the Central Region as part of a contract with the World Bank.

2   Compare Radcliffe and Laurie (this volume) on imaginative geographies of policy.

8 On the spatial limits of culture in high-tech regional economic development: lessons from Salt Lake City, Utah

Al James

Introduction: Silicon envy and the copycat cluster league

Over the last decade the connections between innovation and regional growth have become a central avenue of research for academics and policy-makers alike. With the shift to a new phase of capitalism (variously labeled the information economy, postindustrialism, the Fifth Kondratiev, post-Fordism, or the New Economy) the most important type of competitiveness is argued to be sustained only through learning and innovation, to anticipate and outrun attempts at imitation by competitors. Geographers have argued that these key processes are territorially embedded, and examined how *regional* industrial "clusters" foster and support conditions conducive to knowledge creation, inventiveness, and learning. No region has been more intensely scrutinized than Silicon Valley, California. Widely regarded as *the* paradigmatic icon of economic and technological success in the global knowledge economy, this region is consistently held up as a visible example of what other regions can do to distinguish themselves in a world of heightened spatial competition and of eroding regional policy (Markusen 1999). Consequently, governments across the world have become fixated with trying to recreate the conditions that engendered the region's vitality and innovation, in order to grow their own high-tech "clusters."

The copycat cluster league now includes a Silicon Plateau, Silicon Desert, Silicon Hollow, Silicon Alps, two Silicon Vineyards, five

Silicon Islands, and the improbable Silicon Glacier and Silicon Sandbar.[1] However, beyond clever names and promotional websites these policy attempts to emulate Silicon Valley's success have typically failed to ignite any meaningful long-run processes of regional economic growth (Florida and Kenney 1990; Castells and Hall 1994; Asheim and Cooke 1999; Markusen 1999). Problematically, while policy-makers have sought to recreate the Valley's success through installing the "right" mix of sleek low-rise glassy office buildings, broadband internet connections, a few multinational companies, government-run research institutes, venture capital, reduced taxes, and an appropriately skilled workforce, merely providing an abundance of these "hard" institutions does not "automatically" generate the crucial interactions between firms that underpin the technological spillovers and regional dynamism (Massey, Quintas, and Wield 1992; Johannisson, Alexanderson, Nowicki, and Senneseth 1994; Hudson 1999; Malecki and Oinas 1999). Rather, this is increasingly recognized as dependent on a set of intangible cultural conventions, norms, values, and traditions that facilitate and regulate the economic behavior of firms and that hold these regions together beyond mere economic specialization and interlinkage.

Nevertheless, despite this growing consensus, the nature of this regional cultural "glue" has yet to be fully elucidated, with references to culture still seen by many as appealing to a set of "mystical forces" or a "dustbin category" into which we lump anything sociological that we cannot fully explain (Sayer and Walker 1992; Gertler 2004). In this chapter I seek, therefore, to extend our understanding of the cultural economy of high-tech regional industrial systems, examining: (i) *how* regional culture shapes and conditions firms' innovative behavior and economic performance; and (ii) how far we might empirically verify those links. My analysis focuses on the high-tech industrial agglomeration in Salt Lake City, Utah, embedded in a highly visible regional culture, Mormonism. First, I sketch the evolution of the main schools of thought through which geographers have sought to theorize high-growth regional economies over the last two decades. Second, I show how the cultural economy of firms in the region can be understood in terms of a series of tensions between self-identified regional cultural traits also identifiable within local firms, versus key elements of corporate cultures consistently shown in the regional learning literature as positively underpinning firms' abilities to innovate. Third, I measure the impact of those tensions on firms' economic performance across a series of metrics. Fourth, I outline the main mechanisms and causal agents through which culture shapes and conditions the behavior of firms in the region. Finally, I return to the wider policy relevance of my analysis, arguing for an explicit recognition of the spatial limits of cultural economy on high-tech cluster policy.

Theorizing dynamic regional economies: from transaction cost reductions to cultural economy

Received wisdom has long held culture and economy as discrete entities, each with its own set of self-determining institutions, rationalities, and conditions of existence. However, since the early 1990s economic geography has witnessed a "cultural turn" in which culture and economy are reconceptualized as existing in a two-way recursive relation, based on their perpetual and simultaneous (re)construction by human agents whose economic motives and logics are inseparable from their own sociocultural identities (Crang 1997; Sayer 1997; Barnett 1998). On the one hand, this rethinking represents a response to a series of new economic empirical realities, in which the social bases of economic success (and failure) have become increasingly apparent. On the other hand, it also represents an ontological backlash against the economism of previous Marxist and neoclassical accounts of geographical change. Thus, under the new banner of "cultural economy," scholars have brought to the fore so-called "background" factors: the "soft" sociocultural aspects of economic behavior previously ignored in conventional economic analyses (Wolfe and Gertler 2001). That is, they have sought to put back that which should never have been taken out (see Amin and Thrift 2004: xiv).

This profound shift has been especially apparent within the industrial district literature, as geographers have sought to understand the workings of the post-Fordist spearheads of regional economic growth.[2] Three main schools of thought are discernible. Beginning in the 1980s the "Californian school" sought to explain the growth of highly competitive regional networks of firms across Western Europe and North America through an appeal to the transaction cost reductions gained by firms as a function of spatial agglomeration. First, instability of markets and an accelerated pace of technological change are argued to be met by a disintegration of the production process, and a deepened social division of labor, allowing more flexibility (Scott 1986, 1988). Second, as inter-firm transactions become more important, so the costs of "transport, communication, information exchange, searching and scanning" create a "spatial pull" to minimize these transaction costs through external economies of scale (Scott 1988: 176). Third, spatial agglomeration is also argued to create specialized institutions, which further lower transaction costs as labor and other factors of production are pooled among large numbers of specialist producers.

From the late 1980s, however, the transactions cost approach received significant critique. Problematically, external economies and cost-price benefits to producers cannot fully explain the *ongoing* innovative capacity of particular high-tech regional economies, nor why some remain competitive while others do not (Storper 1997). Saxenian's (1994) work on the divergent regional economic trajectories of Silicon Valley and Boston's Route 128 through the 1980s demonstrated this most explicitly. Controlling for industrial sector, products, historical period, position in the business cycle, political events, and nation-state (Gertler 1995), Saxenian instead highlighted the importance of local *cultural* determinants of industrial adaptation. In Silicon Valley, a distinctive regional culture characterized by a willingness to embrace risk, and loyalties to transcendent technologies over individual firms, underpinned a regional network-based industrial system based on blurred inter-firm boundaries and flexible adjustment among producers of complex related products.[3] In contrast, a traditional conservative East Coast business culture in Route 128 sustained relatively integrated corporations, lesser interaction, and lower rates of economic growth.

Subsequent scholars have built upon Saxenian's work as part of a broad "sociological institutionalist" school. While not denying the reality of transaction cost reductions, this second approach emphasizes that regional economies are integrated by a set of less tangible but crucially important "untraded interdependencies" (Storper 1995) which go beyond simple input–output links. The most common framework employed here has been a geographical application of the Polanyian notion of "embeddedness,"[4] which argues that all economic relationships are *embedded* in both economic *and* non-economic institutions which create distinctive patterns of constraints and incentives (Polanyi 1944). Accordingly, scholars have examined how culture – in the form of collective beliefs, values, norms, conventions, ideologies, taken-for-granted assumptions, and lifestyles – shapes firms' economic strategies and goals, and in particular, how it helps create a sense of community and trust upon which inter-firm cooperation is predicated in an uncertain post-Fordist climate of demand (e.g. Lorenz 1992; Gertler 1995).

In the third variant of regional theory developed largely over the last decade, scholars have extended the sociological institutionalist school to focus explicitly on the significance of knowledge production in understanding regional economic development. This literature is now vast (see Mackinnon, Cumbers, and Chapman 2002 for a useful recent review). However, at the broadest level the advantages of agglomeration are argued to emerge from localized information flows, technological spillovers, collective learning, and the creation of specialized pools of knowledge and skill. These are premised in turn on *formal* and *informal* networks of interaction between firms and their employees which aid the circulation of tacit knowledge within the region (Malmberg and Maskell 1997; Capello 1999). Significantly, scholars have also focused on the qualitative rules, conventions, and norms on which actors draw to combine varied skills, competencies, and ideas to create new knowledge and so underpin innovation. Innovation is therefore increasingly regarded as a fundamentally interactive, and hence unavoidably sociocultural, process (Malecki and Oinas 1999; Asheim 2001).

Unpacking and repacking the fuzzy cultural dustbin

However, while there is a growing consensus that distinctive regional cultures play a vital role in facilitating learning and innovation processes, the cultural economy of innovative industrial systems remains only partially specified within the geographical literature. Problematically, accounts return again and again to descriptions of the cultural properties of these regions, yet rarely specify the exact nature of the causal mechanisms by which regional culture structures promote innovative activity more successfully in some regions than in others (Asheim 1996; Storper 1997). There is often circularity of argument; that innovation occurs because of the presence of certain cultural institutions, and that those cultural institutions are what exist in regions where there is innovation. Indeed, while it is Saxenian who takes us furthest away from this unsatisfactory state of affairs, even she does not thoroughly establish the causal link between the competitive culture she describes and the success of Silicon Valley as a regional economy (Markusen 1999: 879).

To begin to demystify these links, I draw on the multi-tiered conception of culture as has been operationalized in the organizational studies literature. Here (corporate) cultures are conceptualized as the sets of social conventions, embracing behavioral norms, standards, customs, and the "rules of the game" that underlie social interactions within the firm. These conventions are in turn linked to a deeper set of underlying core values that provide more general guidance in shaping behavior patterns within the firm. Corporate cultures are thus seen as coherent and unifying systems that ensure stability and the smooth running of organizations through defining appropriate ways of behaving, attitudes, and ways of thinking, or put more simply, "the way we do things around here" (Deal and Kennedy 2000: 4). Importantly, I argue that we might also apply this model at the regional scale, while simultaneously recognizing that corporate cultures and regional cultures do not exist in isolation from each other; rather, they are interwoven across scales. Thus, rather than the all-encompassing notions of "regional culture" often employed in the regional learning and innovation literature, I instead argue for recognition of a regional culture scalar hierarchy, made up of: (i) individual corporate cultures; (ii) a regional industrial culture; and (iii) the broader regional culture in which these are set (see also Oinas 1995).

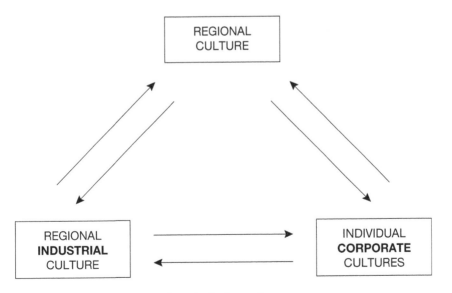

Figure 8.1 Hypothesized culture hierarchy in the region.

Once we unpack culture in this way (see Figure 8.1), the cultural economy of firms in the region can be understood in terms of the *overlaps* between the different levels of this hierarchy. That is, in terms of *regional* cultural systems of collective beliefs, ideologies, understandings, and conventions being manifest in firms' corporate cultures, and hence shaping *firms'* systems of organizational control, rule systems, and decision-making processes. As such, patterns of corporate behavior become regionally culturally-inflected, and it is the various manifestations of these overlaps, their material impacts on firms' economic performances, and their driving mechanisms and responsible agents which form the focus of this chapter. My analyses draw on the case study of the high-tech industrial agglomeration centered on Salt Lake City, Utah, which is inseparable from a distinctive and highly visible regional culture, Mormonism.

Salt Lake City: high tech meets Mormonism

Salt Lake City is the main concentration of population on Utah's Wasatch Front, an urban corridor made up of four counties running north and south for 85 miles along the base of the Wasatch Range of the Rocky Mountains to the east of Salt Lake City. This region is home to more than three-quarters of Utah's population of 2.13 million, along with over 2,100 high-tech firms (90 percent of Utah's total high-tech industry), which currently employ over 70,000 people across a range of subsectors (Table 8.1). Computer Software (SIC 737) is Utah's lead high-tech subsector in terms of both employment and number of establishments, and as such forms my case study industry (following Markusen 1994). Utah's computer software industry contributes over 45 percent of the state's total high-tech payroll of US$1.4 billion (Utah Department of Workforce Services 2001).

Importantly, the Wasatch Front is also the heartland of Mormonism, the regional culture from which the historical development of Utah's high-tech industry is inseparable. Mormonism is the distinctive culture associated with the "Mormon Church," or more properly "The Church of Jesus Christ of Latter-day Saints" (the LDS Church). Mormonism is especially strong within Utah reflecting the state's position as the geographical, political, administrative, and historical heart of the LDS Church, with Mormons comprising over 75 percent of Utah's total population (Church of LDS/Deseret News 2000), the

Table 8.1 *Utah's high-tech subsector in 2000 (adapted from James 2005)*

SIC	Description	Establishments	Employed	% UT high- tech emp.	LQ
283	Drugs	53	3,998	5.7	0.18
357	Computer and office equipment	28	4,057	5.8	1.51
366	Communications equipment	25	2,953	4.2	1.50
367	Electronic components and accessories	58	3,993	5.7	0.83
371	Motor vehicles and equipment	40	7,904	11.3	1.08
372	Aircraft and parts	40	2,744	3.9	0.83
376	Guided missiles, space vehicles and parts	10	5,342	7.6	0.84
381	Search and navigation equipment	3	645	0.9	0.56
382	Measuring and controlling devices	39	1,028	1.5	0.49
384	Medical instruments and supplies	71	8,383	11.9	4.07
737	**Computer software and data processing services**	**1,438**	**23,042**	**32.8**	**5.76**
873	Research and testing services	335	6,168	8.8	0.25
TOTAL		2,140	70,257	100	–

Source: Utah Department of Workforce Services, 2000.

Note: A location quotient is a measure of spatial concentration: a ratio of ratios. The numerator is the total employment for a particular industry in Utah divided by total state employment. The denominator is the US employment for that industry divided by total US employment. Any location quotient over 1 signifies a concentration above the expected state average, e.g. if is 3, means that there are three times more jobs in a particular industry than there would have been if Utah had had a proportional share of national employment in that sector.

same population from which Utah's high-tech workforce is drawn. Mormon culture is conservative by popular standards with strong family and community impulses (May 2001). It includes prohibitions against alcohol and drug use, a commitment to fasting and prayer, modesty in dress, an emphasis on family and obedience to parents, and concerns for the elderly and the poor. The church also emphasizes the Protestant ethics of diligence, education, and the attainment of skills (Cornwall 2001). Three elements of Mormon culture make it especially suited to my study. First, Mormonism is more than simply a creedal faith; it is a whole way of life requiring an almost *total* commitment in customs, values, and lifestyle (see Kotkin 1993). Second, the demographic dominance of Mormons in Utah creates the possibility of a denomination-specific domination of Utah's *general* culture. Third, Mormonism provides a regional culture whose central tenets are easily articulated and well known, and whose ideologies are written down and easily accessible.

As such, Utah's Mormon regional culture is especially *visible* and therefore allowed me to observe and measure the ways in which culture shapes and conditions the behavior of firms in a high-tech regional industrial system.

The data on which this chapter is based was collected by means of a multi-method approach. First, I sought to identify broad regional patterns through a survey of the lead 10 percent of computer software firms (by 2000 revenue from local Utah operations only) across the four counties of the Wasatch Front.[5] I also categorized these data across "Mormon" versus "non-Mormon" versus "intermediate Mormon" firms, defined in terms of the proportion of a firm's founding and management team who were active Mormons (Table 8.2).[6] I achieved an overall response rate of just over 50 percent, and as such the survey dataset covers the top 20 percent of software firms on the Wasatch Front by 2000 revenue. The firms in my survey dataset employ 7,585 people in Utah, and in 2000 generated a combined revenue of US$1,031 million from their Utah operations.

Table 8.2 *Basic distribution of the survey and case study samples (adapted from James 2005)*

Firm Type	Survey				Case Study
	Micro (1–19 emp)	Medium (20–99 emp)	Med-Large (100–499 emp)	Total	ALL 20–99 emp
MORMON majority founded and managed	18	28	12	58 (55.2%)	6 (30%)
INTERMEDIATE Mormon/ non-Mormon mix of founding and management	6	6	9	21 (20.0%)	8 (40%)
NON-MORMON majority founded and managed	10	12	4	26 (24.8)	6 (30%)
Total Firms	34 (32.4%)	46 (43.8%)	25 (23.8%)	105 (100%)	20 (100%)

Note: Of the 105 firms in the survey sample, 89% were founded in the State of Utah, and the remaining 11% typically branch offices of computer software firms headquartered in other US states, especially California. Due to the relatively small number of non-Mormon firms in the medium-large size category (N = 4), I did not include *statistical* comparisons between Mormon and non-Mormon firms in this size category in my subsequent analysis.

Second, I sought to explain patterns manifest in the survey data through a series of in-depth case studies based on semi-structured interviews.[7] In total, 20 broadly similar computer software firms were chosen in order that they cover the spectrum of Mormon and non-Mormon founding and management, as outlined in Table 8.2. In this case study sample,[8] I was also able to expand my initial definition of Mormon firms also to include the *proportion of total* Utah employees in the firm that are active Mormons. In 2000 the firms in my case study sample employed 1,009 people in Utah and their Utah operations generated a combined revenue of over US$111.3 million. All of these firms also fall in the 20–99 employees category, the dominant size category in the survey sample. Finally, to elucidate the processes through which the innovative behavior and economic performance of firms are regionally culturally inflected I undertook a systematic analysis of the interview transcripts through a process of progressive qualitative hypothesis testing.[9]

Unpacking the Mormon cultural conditioning of firms' innovative capacities

At the broadest level, the impact of Mormonism on Utah's computer software industry is manifested through large numbers of Mormons founding and managing Utah's lead software firms. Almost three-quarters (69 percent) of the software firms in my survey sample are Mormon founded; 68 percent have a Mormon majority management team; and 58 percent are both Mormon founded *and* managed (James 2003). Mormons also populate firms' workforces at lower levels, comprising approximately 69 percent of firms' *total* employees in the in-depth case study sample. However, how much difference do Mormon founders, management teams, and majority workforces actually make to the ways these firms operate? The data in Table 8.3 show how this impact can be understood in terms of a series of sustained tensions, between self-identified Mormon cultural traits also manifest within local firms (LHS), versus key elements of corporate and industrial cultures that have been consistently shown in the regional learning literature as positively underpinning firms' abilities to innovate (RHS).

In the following four sections I show how, in some cases, these tensions reinforce firms' innovative capacities; in other cases they potentially constrain them. However, I want to stress from the outset

Table 8.3 *Unpacking the cultural economy of computer software firms on Utah's Wasatch Front with regard to firms' innovative capacities*

Regional Mormon Culture (Self-identified and self-evident)	Regional Industrial/Corporate Culture (Identified in the regional learning and innovation literature as promoting success)
Unity and mutual trust	Interfirm co-operation and trust
Self-sufficiency and autonomy	Use of external competencies
Family (then Church) above all	Sleeping bags under the desk
Respect for established ideas, hierarchy and authority	Creative dissent, multiple advocacy, constant questioning

that this is *not* an anti-Mormon work. Above all, my aim is to foster a better understanding of the processes through which culture and economy are intertwined within firms in the region, *not* to deprecate the LDS Church, its beliefs, or doctrines.

Mormon unity and cultural trust versus inter-firm networking and studied trust

Over the last two decades, the geographical literature has consistently shown that firms' abilities to innovate are strongly conditioned by their access to external sources of knowledge (see e.g. Camagni 1991; Maillat 1995; Oinas and Malecki 2002). When individuals with partially overlapping knowledges come together and seek to articulate their ideas collectively, they are forced to derive more adequate ideas about the technology they are trying to develop (Lawson and Lorenz 1999: 312). Additionally, interaction also provides a basis for comparison of evolving ideas with other practices that are not internally generated. Overall, there results an increased potential for new interpretations and synergies to develop (Capello, 1999). Significantly, the Mormon regional culture on Utah's Wasatch Front is itself characterized by strong ethics of unity, reciprocity, and mutual commitment that shape and condition the nature of interaction among its members (Poll 2001). Indeed, extensive ties of extended Mormon kinship are explicitly cultivated by the LDS Church leadership, induced not only by the belief that unity is a Christian virtue, but also by the trying times experienced by the Mormon pioneers (Arrington and Bitton 1992).[10] Thus, "at the largest level the [Mormon] church as a whole is a family, and . . . Mormons give to others without expectations of recompense, even to those that they do not know" (Dunn 1996: 36).

One way to examine the extent to which local *firms* exhibit these Mormon cultural traits is to track the extent to which Mormon ownership and management affect firms' choice of strategic partners. Significantly, the Mormon founded and managed firms in my case study sample *do* have a higher proportion of strategic partners in Utah who are similarly Mormon founded and managed (67.5 percent) than do their non-Mormon (50 percent) and intermediate Mormon counterparts (54 percent). The rationales driving these patterns have a cultural-historical basis:

> Mormons in the business world in Utah are very protective of each other. When we first settled Utah it was an "us-versus-them" mentality for the first hundred years or more and that mentality of us versus the rest of the Gentile world still translates into business – there's a lot of cooperation between Mormon high tech businesses.
>
> (Creative Director, active Mormon male)

> Mormons go to Mormons. I can't tell you how many times in meetings I've heard "Oh he's in my Ward." And as soon as they discover I'm not a Mormon there's a barrier goes up and I have to establish a level of trust that would automatically be assumed if I was Mormon.
>
> (Software Consultant, non-Mormon male)

These distinctive Mormon–Mormon network structures therefore highlight the importance of active social connections (or "social capital") in shaping the quality and quantity of interactions among Mormons, premised on mutual understanding, norms of reciprocity, and trust (also Cohen and Prusak 2001; Putnam 1993; Porter and Lyon, this volume). Trust[11] is critical in enabling inter-firm interactions, due to frequent and unanticipated contingencies associated with technical change or demand shifts, both of which provide opportunities for firms to interpret contracts in ways that shift the distribution of returns to the favor of one side (Lorenz 1992: 199). However, in contrast to dominant characterizations of ("studied") trust as something which emerges over time based on firms' judgment of others' behavior in repeated encounters (see e.g. Sabel 1992; Lazaric and Lorenz 1997), the Mormon firms in my study have a propensity to trust each other in the *absence* of repeated interaction. I label this "cultural trust," premised on a common history and heritage, and belief in the same God. As part of this, general teaching, preaching, and organizational roles within the LDS Church ("callings") that connote good standing within the general Mormon community (Shepherd and Shepherd 1994) are also invoked in Utah's computer software industry, often forming the basis for *explicit* corporate decisions:

> There's certainly that thing where you'll run across another company and the founder is Mormon as well, and you kinda think "let's pay this guy a little more attention." You'll give them a little more, because there's that greater sense of he is what he says he is.
>
> (CEO and Founder, active Mormon male)

> I try to make sure that anyone we work with is active and not just "nominally," but that they're really living the principles of the Gospel of Jesus Christ. Why? – Because the chances are they'll be honest and have integrity in our business dealings.
>
> (CEO and Founder, active Mormon)

Mormon customs, conventions, and social norms therefore generate a "cultural closeness" between firms, a social architecture that aids working alliances. However, "the obvious corollary of the blurring of boundaries between self and other in the context of the [LDS] church is that the boundary between the domain of the church and the 'outside world' is drawn at the same time" (Dunn 1996: 38). Thus, while cultural trust in the Utah high-tech business setting helps sustain interaction between *like* firms, it simultaneously *excludes* firms that do not share the same cultural markers, hence constraining firms' abilities to learn from non-like firms. More broadly, this also highlights the weakness of social capital, in which particularist loyalties translate into group exclusivity (see Pieterse, this volume; Portes and Landolt 2000).

Self-sufficiency and autonomy versus outsourcing and use of other firms' competencies

Successful learning and innovation therefore require that firms maintain close networks of external association. Additionally, inter-firm collaboration is an important means of broadening firms' capacities more widely (Fountain 1997; Hotz-Hart 2000). Firms enter alliances to combine their own competencies with those of a partner to create a competitive position that neither could have achieved alone. Thus in the context of increased complexity and intersectoral nature of new technologies, and shortening product life-cycles, partnerships allow firms to speed the pace of product introduction, improve product quality, and move more quickly into new markets (Hutt, Stafford, Walker, and Reingen 2000). In contrast, Mormon culture is instead characterized by strong emphases on individual self-sufficiency, independence, and self-reliance (Ludlow 1992), ethics also manifest in local software firms. These social ideals are explicitly taught by the LDS Church, and have their historical roots in the Mormon pioneer experience when Utah's hostile physical environment forced Mormon families to hone the virtue of self-sufficiency in order to survive (Young 1996).

My results suggest that software firms on Utah's Wasatch Front exhibit Mormon cultural emphases on self-sufficiency through their *overall* levels of strategic partnering. Specifically, the Mormon founded and managed firms in my survey and case study samples have on average around half as many strategic partner firms as their non-Mormon counterparts in each firm size category.[12] Thus, while Mormon firms have a higher propensity to interact with other Mormon firms, their *overall* levels of inter-firm networking are reduced relative to non-Mormon firms:

> We run into this all the time, I call it the "Pioneer Spirit." There tends to be this feeling of well we've got it covered, we can do it ourselves, and so there's a lesser willingness to bring in outsiders, to farm it out. It's an approach to business that is very reflective of the Mormon historic reality.
>
> (Vice President and Co-Founder, Mormon female)

> That insularity is very much reflected in Mormon business approaches, there's a real tendency to keep everything in-house. So we've never looked out of the window, we've never even looked at our competitors' products, at anybody else's ideas.
>
> (Director of Corporate Research, active Mormon [convert] male)

While I found typical pride among respondents in their firms' self-sufficiency and having grown something from nothing, many firms were also aware of the limits of such an introverted approach:

> I've seen a lot of Utah companies really hurt themselves because they've said "Oh I don't need an attorney," or "I don't need an accountant, I can structure this loan or investment myself." And that makes it difficult for the business to grow and to be successful.
>
> (Chief Financial Officer, active Mormon male)

> In the first incarnation of [the company] it was that way to a fault. Whenever we needed to mass-produce, rather than outsourcing it we

went out and bought our own duplicator. And then when it came
to shrink wrapping, do we farm that out? – No, we go buy a shrink
wrapper and do that ourselves! So we outsourced nothing and got so
bogged down that we kinda delayed getting to our overall goals.

(President and Founder, active Mormon male)

Further, a whole range of studies have demonstrated that where firms
are introverted and rely mainly on internal resources their individual
performance is weakened, along with that of the entire regional
system (MacPherson 1992; Wiig and Wood 1997). Rarely does a
single firm have superior capabilities in all phases of the production
process, and so it is imperative that they take advantage of
the synergies that flow from shared enterprise. As such, the
introvertedness of particular Mormon founded and managed firms
can be viewed as potentially constraining their innovative capacities
by limiting their access to external sources of information and
expertise.

Family (and church) above all versus sleeping bags under the desk

In her analysis of Silicon Valley and Route 128's divergent
performances through the 1980s AnnaLee Saxenian highlights how
Silicon Valley's engineers, often young men without wives or
families, instead developed shared identities around the project of
advancing new technologies (Saxenian 1994). Crucially, this then
facilitated an increased commitment to the firm, a regularity of long
unsocial work hours, and the completion of large workloads in short
periods of calendar time. These are competitive advantages to the
firm particularly when bringing a new product to market. In contrast,
Mormon culture is characterized by a rigid separation of work
and social life through strong commitments to family and church.
Mormonism emphasizes the primacy of procreation and traditional
family values, and as a result its members are characterized by
higher rates of marriage, lower rates of divorce, higher fertility rates,
and larger families than the US national average (Heaton 2001).
Children are also considered of far greater value than any material
wealth, with the favorite dictum of former church President David O.
McKay often quoted in support; that "no success can ever
compensate for failure in the home."

Various sociological studies have demonstrated how these cultural
commitments detract from devoting large amounts of time to careers

(Heaton 2001). Moreover, these cultural priorities are also manifest within computer software firms in Utah, impacting upon levels of work intensity across four areas. First, the Mormon (founded *and* managed) firms in the case study sample have average work weeks that are 5.8 hours (approximately 10 percent) less than those of their non-Mormon counterparts (44.7 hours per week vs. 50.5 hours per week).[13] Second, while extra-long work hours (working beyond 7 p.m.; working through the night; working weekends; and/or working overtime) are a regular occurrence in 67 percent of the non-Mormon founded and managed firms and in 62 percent of the Mormon Intermediate firms in any particular month, this compares with only 42 percent for their Mormon counterparts. Indeed, in the majority of Mormon managed firms this is a *deliberate* strategy:

> Family and church are very high priorities for us, so we try not to have our employees working a lot of extra hours, and we try to reduce that as much as we can in order for them to participate in religious activities and family activities which we see as a very important part of our culture.
>
> (Director of Business Development,
> active Mormon male)

Third, the most frequently cited area of extra-long work hours involved Sunday working, a major tension for Mormons for whom the Sabbath has been historically set apart as a holy day of worship, for rest and spiritual renewal. Thus, while two-thirds of the Mormon firms in my case study sample have company policies that restrict Sunday working, none of their non-Mormon or Intermediate Mormon counterparts maintain such policies. This is significant given the 15-month development schedules common among the other firms in my case study sample. Typically, in the final stages of a software project, weekend working and late evenings are the norm in the non-Mormon firms:

> The last two or three months of a project, you're working 80 hours weeks, and those last few weeks get insane! On my last project, in the last month I slept in my own bed twice! We got sofa beds in the basement and so you'd just go down there and grab a few hours and carry on – insane! But good employees, Mormon or not, they know that's what this industry's about, and they know they get paid better because of it.
>
> (Chief Financial Officer, non-Mormon male)

These attitudes to Sunday working contrast with those that characterize the corporate cultures of the majority of the Mormon firms in my case study sample:

> Even when we've had 24 hour a day 6 days a week operations, we always close Sunday! I've just always tried to maintain that. We don't want work replacing family or religious structure. Work is supplementary and there to support family and church. True happiness is not achieved through a bigger bonus.
>
> (Director of Marketing, active Mormon male)

Fourth, average lengths of paid vacation are on average six days (approximately 55 percent) longer in the Mormon firms than in the non-Mormon firms (17.5, cf. 11.25 days).[14] Removing the start-ups from the analysis has a negligible effect on the data (18.5, cf. 11.5 days). Further, while holiday lengths are typically closer to the European average of four weeks in the Mormon firms, in the non-Mormon firms they are closer to the US average.[15]

Overall, my results suggest that Mormons in Utah's computer software industry have created a corporate environment in which work is valued differently than in typical US software firms. Mormon cultural emphases on the importance of family, a balanced lifestyle, that no success can ever compensate for failure in the home, and of money as merely a means to a higher spiritual end all impact upon the structure of firms' work weeks and mean that extra-long working hours are discouraged. Many respondents outlined the various constraints that they perceived these culturally informed work patterns to have on their respective firms:

> There very definitely is a strong sense in the Mormon Church that your job and your success are not the most important thing in your life. So a lot of times I wonder if we're as effective at competing, precisely because we can't, and we won't, throw everything we've got into the company. So you've got firms elsewhere working 80, 90, 100 hour work weeks, often putting in twice as many hours as we did in the same week!
>
> (Director of Technology and Co-Founder, active Mormon male)

> We continually battle the problem of our managers from other US locations feeling like the workforce in Utah aren't as committed. And that is reflected in lower levels of new product introduction in the Utah office. If we have a big product release scheduled, your Utah

employees are more likely to say "well we'll just have to push it back and reschedule." Whereas in some of our other office locations, there's more of a willingness to spend 24 hours a day at work until it's done.

(Director of Human Resources,
active Mormon female)

Indeed, these self-perceived constraints are supported by secondary analyses at the US national level. Notably, a 1995 report by the US National Center for the Educational Quality of the Workforce (EQW)[16] investigated the relationship between firms' work hours and their overall productivity, estimating that for a 10 percent increase in work hours in non-manufacturing firms, there results a 6.3 percent increase in establishment productivity[17] – significantly, my results suggest that the non-Mormon firms work 16 percent longer per week than their Mormon counterparts (50.0 hrs/week vs. 43.0 hrs/week). However, respondents also highlighted a number of perceived advantages that accrue to the Mormon-founded and managed firms from these culturally imposed limits on work hours, the most commonly cited advantages centering on a more healthy, less stressed, and consistently productive workforce:

It's true that you can work 18 hour days for me and probably produce a lot, but not over the long-term. So the church's emphasis on balance and family and taking care of yourself results in a more consistently productive workforce. It's not just about producing more, but about producing better. That's the strategic advantage.

(Director of Technology Development,
active Mormon)

In examining the impact of Mormon cultural economy on the competitiveness of software firms on Utah's Wasatch Front, it is therefore imperative that we not view the activities of production in isolation from those of *re*production, and hence the social sustainability of culturally informed work practices. I return to this point later in the chapter as part of a broader discussion regarding alternative metrics of firm competitiveness.

Respect for authority, established ideas, and group unity versus creative dissent and questioning of leaders

Continuous technological learning and innovation are therefore highly dependent on firms' abilities to *access* external sources of information and knowledge. Fundamentally, however, they are also dependent on firms' abilities to assimilate, reconfigure, transform, and apply new information to commercial ends; that is, on firms' "absorptive capacities" (Cohen and Levinthal 1990). Different absorption rates are not random but depend on the social and cultural structures within firms, because the ability to absorb new knowledge will always depend on sociocultural constructions of what is acceptable and desirable (Schoenberger 1997; Westwood and Low 2003).[18] The organizational learning and innovation literature has consistently highlighted a set of cultural norms that, if widely shared by the members of a firm, actively promote the generation of new ideas and help in the implementation of new approaches. These include a climate of openness in which debate and conflict are encouraged; a willingness to break with convention; widespread support for trying new things; the right of employees to challenge the status quo; and multiple advocacy, that learning requires more than one "champion" if it is to succeed (DiBella, Nevis, and Gould 1996; Deal and Kennedy 2000). Firms' abilities to innovate therefore presume a necessary relationship between learning and active employee involvement at all levels; that all employees can act as independent agents, take responsibility, experiment, and make mistakes as they learn (Spender 1996).

However, these traits contrast with Mormonism, from which the development of Utah's software industry has been inseparable. First, Mormon culture is characterized by cultural emphases on unity and individual sacrifice for the common good, which sustain strong tendencies towards conformity within the group (Shupe 1992). Second, these are reinforced by a pervasive respect for established ideas and church operating procedures (Ostling and Ostling 1999). Third, the LDS Church organizational system is also based on predominantly top-down flows of information, in which leadership decision are never challenged, only supported by the wider Mormon populace. Indeed, an oft-cited dictum used to justify this system is that "when the Prophet speaks the thinking has been done" (Ludlow 1992). Fourth, these cultural emphases of reverence for established ideas and leadership authority are in turn reinforced by wider Mormon emphases on being passive, non-confrontational and never demeaning another person.

Significantly, this distinctive set of Mormon cultural traits is also manifest in the corporate cultures of firms in my case study sample. Approximately 40 percent of these firms are internally perceived by the industry respondents as having corporate cultures that place a premium on unity within the firm, and a "follow thy leader" mentality. Interestingly, this includes two-thirds of the Mormon firms and half of the Mormon intermediate firms, but only one of the non-Mormon firms:

> Unity is a very much higher value to us as Latter-day Saints. We don't have quite the intrinsic treasuring of diversity that I see among my colleagues who aren't Mormon. All the time, no-one ever comes out with an opinion, only collective opinions.
> (Director of Research, active Mormon male)

> In the church, when we have a new leader appointed, everybody raises their hand to show that they support that person in their calling. And I've noticed that kind of mentality within this company: we understand there is one leader in charge, and that we support his opinions and decisions in the same way we would in a church setting.
> (Production Manager, active Mormon male)

Indeed, over half of the non-Mormon industry respondents (spread across the whole spectrum of Mormon through non-Mormon firms) identified their Mormon colleagues and employees as *generally* less willing to question ideas and leadership authority. Additionally, almost one-third of the (47) active Mormon industry respondents also identified this trend among their fellow Mormon employees and colleagues generally, arguing that Mormon managers and employees raised in the Church simply "borrow" from the models that are familiar to them.

Respondents outlined a number of ways in which these Mormon-inflected corporate cultures offer advantages to the firm. First, respondents suggested that unity through a common value base makes it easier for the firm to mesh as a team, consistent with norms highlighted in the innovation literature as promoting the implementation of new ideas in the firm. These include teamwork, a shared vision, and a common direction upon which firms can build consensus, mutual respect, and trust (O'Reilly 1989). Second, I found a widespread appreciation among both Mormon and non-Mormon respondents alike of the more friendly and less stressful work environments that these Mormon-informed corporate cultures sustain:

> Mormon culture teaches us to be more passive and less confrontational, and I've always found a much more mild atmosphere

> in Utah offices than in other places of the US I've worked before. You
> don't have the passionate, loud, angry discussions that I've seen in
> other organizations where people sometimes even end up leaving
> because of that. People are just a lot more supportive of each other
> here in Utah.
>
> (Director of Human Resources,
> active Mormon male)

However, respondents also identified a number of disadvantages of
these distinctive Mormon-inflected corporate cultures, in terms of
Mormon cultural traits of respect for established ideas, unity, and
top-down leadership authority undermining the processes of creative
dissent, constant questioning, and multi-directional knowledge
flows that underpin innovation in firms.[19] Indeed, the majority of
the Mormon industry respondents were themselves aware of these
limits:

> It's a strange paradox to be working in high tech in this Mormon
> environment where obedience is expected and questioning authority is
> frowned upon in a very direct way. So you often get R&D situations
> where people are just doing that they're told rather than thinking
> outside the box and helping the company come up with the creative
> solutions that they need to beat their competitors. But if you don't
> question you will never survive in this business.
>
> (Director of Research, active Mormon female)

> Sometimes there's too much of a lack of confrontation among Utah
> employees, where they don't speak up when they should, they don't
> raise issues because they're afraid of creating contention and hurting
> other's feelings, and I think that makes us less effective.
>
> (Director of Human Resources,
> active Mormon male)

As such, the prevalence of Mormon cultural emphases on unity and
strong conformity within the group, respect for established ideas and
leadership authority, and passive non-confrontation can be interpreted
as further manifestations of the cultural economy of Utah's high-tech
cluster, which shape and condition the absorptive (and hence
innovative) capacities of local firms. How far we might measure the
material impact of this set of cultural traits upon local software firms,
along with the others outlined in this section, forms the focus of the
next section.

Measuring the impact of (Mormon) culture on firms' economic performance

While the regional learning and innovation literature has reached a consensus that culture is critical for understanding the dynamics of regions, the precise nature of *how* regional culture shapes and conditions the innovative behavior and economic performance of firms has yet to be fully elucidated. I argue that the cultural economy of software firms on Utah's Wasatch Front is best understood in terms of various Mormon cultural conventions, norms, values, and beliefs being manifest in local Mormon founded and managed firms, conditioning in turn their decision-making processes and hence observed patterns of behavior. Moreover, there exists patterning at three levels. First, patterning is apparent *within* firms across the different cultural traits outlined in the previous sections. Second, this patterning is in turn consistent with *regional* cultural values. Third, there exists patterning *across* different firms' corporate cultures, sustaining a regional *industrial* culture (see Figure 8.1). In some cases, the prevalence of regional Mormon cultural traits within firms' individual corporate cultures potentially enhances and reinforces their innovative capacities; in other cases it potentially constrains them. To measure the *overall* impact of these tensions on firms' economic performance, I have employed five metrics of competitiveness: (i) linear revenue growth since start-up; (ii) assumed exponential revenue growth since start-up; (iii) R&D intensity I (R&D expenditure to annual revenue); (iv) R&D intensity II (R&D employment to total employment); and (v) productivity in terms of revenue per employee. The results are shown in Table 8.4.

Overall, the data in Table 8.4 show that for four of the five metrics of firms' economic performance, the non-Mormon firms outperform their Mormon counterparts, as highlighted in bold. Significantly, the age distributions of the Mormon and non-Mormon firms are almost identical for each of the employee-size categories employed. While limits of space preclude a step-by-step analytical discussion here (see James 2005), the most striking differences in economic performance at the survey level include: exponential growth rates, where the non-Mormon firms outperform their Mormon counterparts three times over (micro category); Type I R&D intensities (non-Mormon firms, medium size category, two times greater): and productivity (non-Mormon firms, micro size category, over two times greater). At the case study level, the most striking differences include: linear growth

Table 8.4 *Measuring the economic performance of Mormon versus non-Mormon computer software firms on Utah's Wasatch Front (from James 2003)*

Metric of Firm Competitiveness	Survey Sample (105 firms)				Case Study Sample (20 firms)	
	Micro (1–19 emp.)		Medium (20–99 emp.)		(20–99 emp.)	
	Mormon	Non	Mormon	Non	Mormon	Non
(i) Linear revenue growth since start-up (2000 UT revenue/age)	0.16	0.32	0.78	1.05	0.18	0.73
(ii) Exponential revenue growth since start-up (2000 UT revenue/age)	0.28	1.05	1.70	1.68	0.56	1.57
(iii) R&D intensity type I (R&D expend: sales revenue)	0.23	0.24	0.22	0.53	0.29	0.59
(iv) R&D intensity type II (R&D emp: total emp)	0.55	0.57	0.40	0.58	0.57	0.34
(v) Productivity ($1,000 revenue/ employee)	60.47	155.71	123.69	88.82	88.74	103.83
Definition of Mormon vs Non-Mormon Firms	Founding and management ONLY				Founding, management AND majority workforce	

rates, where the non-Mormon firms outperform their Mormon counterparts four times over; and exponential growth rates (non-Mormon firms three times greater). Thus while the results are not monolithic, they suggest that the Mormon cultural inflection of the corporate cultures of the Mormon founded and managed computer software firms in my survey and case study samples has a constraining effect on their innovative capacities and economic performance.

However, there clearly remains much to be done in refining and supplementing the metrics used here. As part of that, the data in Table 8.4 employ only a narrow economic definition of competitiveness. However, with the shift to the new economy, workers and families are increasingly being challenged in new ways to combine the activities of production and *re*production, work and home, in an attempt to achieve "work/life balance" (McDowell 2004). As such, it imperative that we also include metrics on the *social sustainability* of culturally informed work practices in our analyses of competitiveness. Indeed, it is this key premise that motivates my current research on the role of socially equitable work–life balance policies in underpinning regional economic competitiveness.

From material impacts to causal mechanisms and responsible agents . . .

In the preceding sections, I outlined how the Mormon regional culture on Utah's Wasatch Front shapes and conditions local corporate forms, observed behavior, innovative capacities, and economic performance. In this section, I outline the main mechanisms and human agents whose daily activities serve to reproduce culture and economy as a seamless ensemble within the region, and through which the activities of local firms become *unavoidably* regionally culturally-inflected.

The first mechanism centers on key individuals who occupy positions of power within the firm. Fundamentally, because what the firm understands itself to be is produced through the actions of its employees, the identities and commitments of these key individuals are closely entwined with (although not identical to) *corporate* identities and commitments (Schoenberger 1997). Further, when somebody is employed in a firm, their cultural upbringing comes with them. As such, it is not just a case of employees' personal values but of *regional* cultural values being brought into the firm, informing decision-making processes, corporate strategy, and hence observed patterns of behavior, through definitions of what has value and what does not. Indeed, for the majority of the respondents, the application of their religious values within the workplace was not only regarded as acceptable, but also a "natural" thing for them to do. Given that 68 percent of Utah's lead 10 percent of software firms are Mormon managed, and 58 percent are both Mormon founded *and* managed, this "key individuals" mechanism is very powerful in the regional cultural inflection of local firms' economic activities.

This second mechanism through which regional culture unavoidably shapes the activities of local firms can be labeled the "strength in numbers" mechanism, based on the group ratification of regionally culturally-informed corporate decisions. Fundamentally, culture is a *group* property (Stark 1996) and, as such, what counts in terms of cultural values limiting firm behavior is not only whether the firm's decision-makers embody those values, but whether those values are ratified by the wider work group and accepted by the majority as a valid basis for action (Stark 1996: 164). If most of the firm's employees do *not* share those values, even if individuals do bring regional cultural considerations into corporate decision-making processes, these will rarely strike a responsive chord in most of the others and instead be smothered by group indifference. The strength of this mechanism lies in Mormons comprising on average 69 percent of firms' total workforces on the Wasatch Front.

The third mechanism through which the activities of Utah's software firms are regionally culturally informed centers on the labor market and, at one level, on firms actively seeking employees that match their corporate values. While *Title VII* of the *US Civil Rights Act* (1964) makes it illegal to discriminate in labor recruitment based on assumptions about the abilities of members of certain religious or cultural groups,[20] firms in Utah *do* discriminate on this basis: over one-quarter of the firms in my case study sample admitted making *explicit* requests on the Mormon background of candidates sought to fill a position. On a second level, this is reinforced by *workers* deliberately seeking firms regarded as holding similar values to their own. The main corporate value preferences vocalized by Mormon candidates at job interview revolved around issues of not working Sundays or on violent, sexual, or gambling software content; earning a wage that is large enough for their wife to remain at home and so maintain a traditional Mormon nuclear family; and of working on a product that has obvious social benefit. While these are not exclusively Mormon value preferences, respondents suggested that only Mormon applicants ever vocalize these issues with explicit recourse to religious justifications.

The fourth mechanism centers on various civic institutions that exercise a high degree of social control over their members' sense of identity and patterns of behavior (see Bahr 1994). In the LDS Church members are taught through ongoing teaching programs that they cannot compartmentalize their religious lives. The LDS Church also operates an educational system of seminaries (in high schools) and 1407 institutes at colleges and universities to provide LDS-orientated educational and social programs for students in secular education. Additionally, the mission further reinforces these mechanisms of socialization and social control,[21] and enlists more than 60 percent of young Mormons of eligible age. Importantly, having defended the church and its doctrines for two years, returned Missionaries tend to be more orthodox and active in the church than other members (Vernon 1980). Overall, these various elements of the LDS Church maintain the strength of members' commitment to, and non-compartmentalization of, Mormonism in later life, reinforcing the Mormon cultural inflection of firms' internal structures via the "key individuals" and "strength-in-numbers" mechanisms outlined above.

The final mechanism through which the activities of Utah's software firms are regionally culturally-informed centers on US national legislation. Specifically, the *US Workplace Religious Freedom Act* (1972) amended *Title VII* of the *US Civil Rights Act* (1964) to require employers to make reasonable accommodation for the religious beliefs of employees and prospective employees, unless doing so would "impose an undue hardship," defining religion as "all aspects of religious observance and practice, as well as belief." The *Religious Freedom Restoration Act* (1997) further increased employers' responsibilities to accommodate workers' religious beliefs within the workplace. These two key pieces of legislation therefore reinforce firms' obligations to allow Mormon religious-cultural values to be incorporated into their internal structures.[22]

The complex intertwining between culture and economy within these firms is therefore not pre-given or static, but continually remade over time, via this set of multi-scaled mechanisms which mediate between everyday social practices within the firm and the reproduction of corporate culture structures that mirror key elements of regional culture structures. I am not arguing that regional culture mechanically or rigidly determines worker and firm behavior, but rather that it structures the material and cultural resources that enable and constrain the action of individuals and the firms in which they work.

Policy matters: high-tech clusters and the spatial limits of cultural economy

The wider implications of my research center on industrial cluster policy, with which policy-makers across the globe have become fixated over the last two decades as an important tool for stimulating economic growth. While clusters are argued to occur in many types of industries, high-tech clusters in particular have attracted most policy attention (Swann, Prevezer, and Stout 1998; OECD 1999; Keeble and Wilkinson 2000; Norton 2001), viewed as offering a clean and high-wage mode of economic development capable of high rates of regional growth. Above all, it is the non-mathematical models of local industrial development articulated and popularized by Harvard business economist Michael Porter that have been widely adopted in policy (see e.g. Porter 1990, 1996, 1998). These emphasize the benefits of cooperation between firms, and between firms and other institutions through which knowledge and information are exchanged. Porter also argues that industrial clustering not only fosters economic growth in the regions in which clusters occur, but is also a key source of international competitive advantage for the industries concerned, in turn underpinning *national* economic growth. The cluster concept also sits well with the policy preoccupation with micro-economic supply-side intervention, and with the growing trend towards decentralization of policy responsibility and with an associated emphasis on developing the indigenous potential of localities and regions (see Martin and Sunley 2001; compare Radcliffe and Laurie, this volume). Moreover, Porter's model has a very clear role for government itself in stimulating and supporting the growth of clusters as catalysts or "animateurs" for cooperation between firms, one of a set of tangible institutional "ingredients" deemed critical to the development of other successful clusters (see Table 8.5).

However, for all this international "policy trade," it has become increasingly clear that many cluster policies do not travel well, that innovative regional economies may not simply be "cloned" elsewhere (Saxenian 1989; Florida and Kenney 1990; Lorenz 1992; Malecki and Oinas 1999; Markusen 1999; Scott 2000). I argue that these limits of cluster policy are in large part a function of the overly narrow economic theory upon which these policies are premised. The whole project seems to be one of "add institutions and stir" in some assumed inevitable linear process, copying explicitly from other successful regions such as Silicon Valley in California, Route 128 in Massachusetts, and biotech in the San Francisco Bay area and the New York tri-state area, which are assumed to offer some blueprint manual. There are three interrelated problems with such an approach.

First, while cluster policies are premised on the notion that high-tech growth in particular areas is dependent on the "right" mix of formal institutions deemed necessary for an innovative regional economy, cluster policy initiatives premised on the provision of particular incentives and disincentives are unlikely to work in some simple

Table 8.5 *High-tech cluster policy "shopping list" of necessary institutions*

Institutional component	Description
Strong science base	Leading research organizations: e.g. University departments, govt labs, hospitals/medical schools and charities
	Critical mass of researchers, world leading scientist(s)
Growing company base	Thriving spin-out and start-up companies
	More mature "role model" companies (exporting and with global Presence)
Ability to attract key talent	Critical mass of employment opportunities
	Image/reputation as a (biotechnology) cluster
	Attractive place to live
Availability of finance	Venture capitalists, business angels
Premises and infrastructure	Incubators available close to research organisations
	Premises with labs and flexible leasing arrangements
	Space to expand
	Good transport links: motorways, rail, international airport
Business support services and large companies in related industries	Specialist business, legal, patent, recruitment, property advisers
	Large companies in related sectors
Skilled workforce	High quality skilled workforce, training courses at all levels
Effective networking	Interfirm networks of suppliers and customers
	Extensive forward and backward linkages
	Regional trade associations
	Shared equipment and infrastructure
	Frequent collaborations between government, business, and the independent sector
Supportive policy environment	National and sectoral innovation support policies
	Support from RDAs and other economic development agencies as catalysts for cooperation between firms with complementary skill-sets
	Sympathetic planning authorities

Source: Modified from *Biotechnology Clusters*, Department of Trade and Industry, 1998 (Section 3.2).

Note: US innovation policy recognizes the following components of innovation systems: Governance, Legitimation, Technology Standards, Scientific Research, Financing, Human Resources, Technology Development, Networks and Linkages, Markets.

linear fashion. Spatial proximity of institutions, the first usual indicator of a cluster, does not guarantee interaction or "automatically" generate cooperative interactions between firms widely theorized to underpin technological innovation and knowledge spillovers in the region (Massey, Quintas, and Wield 1992; Storper 1997; Sunley 2000). Thus, for example, the uneven networks of association documented in the Utah case, in which non-like firms are excluded in favour of interactions with firms and workers of a similar cultural background, demonstrate that physical proximity is less important than *cultural* proximity, defined by shared cultural conventions, norms, attitudes, values, and beliefs. To undermine further the assumed linearity of cluster institutional dynamics, both enablers and constraints on firms' innovative capacities stem from the *same* regional culture in which they are embedded.

Second, while Porter's cluster model does have a limited role for culture, its treatment is highly ad hoc, descriptive rather than analytical. Worse still, regional culture is reduced to a mere bolt-on to orthodox economic analysis, and problematically categorized as a stand-alone variable. However, my analysis suggests that culture is not simply another item to add to the high-tech cluster shopping list, but rather a factor that underlies the functioning of *all* the other institutions within dynamic regional economies. It is therefore highly problematic to simply strip the firm's regional cultural context into an assumed independent variable in this manner (see also Porter and Lyon, this volume). Indeed, this problem is further exacerbated by the increased appetite for importing "off-the-peg" policy solutions at the international scale (also see Peck 1999). Policy measures developed within the US national cultural context have been directly applied to the UK, which not only has a different *national* culture, but also within that its own unique mosaic of regional industrial cultures.

Third, a common cluster policy approach has been simply to exhort firms doing business within a national jurisdiction to modify their own behavior (Gertler 1997: 56), so that, for example, inter-firm cooperation or new ways of working become more commonplace. However, such policy proscriptions fail to recognize that new behavior patterns crucially require that firms and their members bend, or indeed break out of, accepted ways of thinking and acting to develop new frames of understanding (Harrison 1994). It is therefore insufficient to construct institutions that promote openness, cooperation, and collaboration at the level of the region, but which do not address questions of transforming firms' internal corporate cultures. Fundamentally therefore, regional economic development presupposes, yet is not reduced to, cultural change (Oinas 1995). However, while the organizational learning and culture change literatures are huge,[23] they are striking in their viewing corporate culture in isolation. Problematically, however, by not recognizing that firms' corporate cultures are but part of a broader regional cultural scalar hierarchy, a whole series of mechanisms are ignored which strengthen firms' corporate cultures and hence the decision-making processes and corporate behavior patterns that they inform.

Conclusion

Despite an emergent consensus that culture is critical for defining and understanding the dynamics of regions, our understanding of these links remains partial. Drawing on the case study of Utah's high-tech cluster this chapter has focused on three main elements. First, the cultural economy of firms in the region can be understood in terms of a series of tensions between self-identified regional cultural traits also identifiable within local firms, versus key elements of corporate and industrial cultures that have consistently been shown in the regional learning literature as positively underpinning firms' innovative capacities. Significantly, these tensions exist with respect to both firms' abilities to access external sources of information, knowledge, and competencies, *and* to use those once they enter the firm. Moreover, both enablers *and* constraints on firms' innovative capacities stem from the same Mormon regional culture. Second, using a series of metrics it is possible to measure the impact of regional culture on firms' economic performance, and that the overall impact of these tensions on Utah's lead software firms is a constraining one. Third, I have outlined some of the tangible mechanisms and causal agents through which regional cultural imperatives *unavoidably* come to inform firms' internal structures, decision-making processes, and hence observed patterns of behavior.

Overall, the findings presented here have a high degree of potential generalizability. While the nature of the Mormon regional culture studied here is unique, arguably the causal mechanisms through which this particular regional culture shapes and conditions the behavior of local firms are merely locally contingent instances of more general processes also operating in other regional economies.[24] To what extent is it possible to manipulate these various mechanisms, and how that varies across different national economies, are key questions for future research. More fundamentally, if these regional culture structures continue to be ignored, then myths will continue to be perpetuated in policy about the replicability of the formal institutional parts of innovative regional economies which do not function independently of regional culture. The overriding question, therefore, is not whether we should focus on the cultural *or* the economic; but of how we do both at the *same* time, coupled with an ongoing interrogation and evaluation of existing policies to reveal their limitations, with the aim of producing more appropriate and effective forms of policy intervention.

Acknowledgments

This research was funded by the Economic and Social Research Council (Award No. R00429934224). I benefited from constructive criticisms on earlier versions of this chapter presented to colleagues at the University of Newcastle-Upon-Tyne (CURDS); Lund University, Sweden; and St John's College, Oxford.

Notes

1 The definitive collection can be found at www.tbtf.com/siliconia.html. At last count there existed 79 appropriations of names beginning with "Silicon" by 105 areas outside Silicon Valley. These names are either promoted by local boosters and/or assigned to an area in a press account.

2 Post-Fordism refers to the collection of workplace practices, modes of industrial organization, and institutional forms identified with the period since the mid-1970s in advanced capitalist economies, following the era known as Fordism (Johnston, Gregory, Pratt, and Watts 2000: 615). It is characterized by the application of production methods considered to be more flexible than those of the Fordist era.

3 Saxenian's (1994) account has been contested by others; see Florida and Kenney (1990) and Harrison (1994).

4 Polanyi's work on embeddedness was subsequently reintroduced to current debate by Granovetter (1985).

5 This survey (105 firms in total) generated data at the regional level across five key characteristics of firms: (i) basic characteristics in terms of employment, age, location, etc.; (ii) inter-firm relationships and external orientation; (iii) financing histories; (iv) in-house technological capabilities and innovative

processes (occupational structure, R&D employment and expenditure, and R&D intensity); and (v) competitive "performance," in terms of revenue, rates of revenue growth since start-up, and employment.

6   Indeed, this was often proudly displayed in the executive biographies on company websites, and Brigham Young University (BYU) alumni status was a convenient proxy – over 99 percent of BYU students are active members of the LDS Church and a BYU education has been consistently shown to have a strong positive impact upon LDS commitment in later life.

7   The interviews were conducted over a five-month period (May–September 2001) with employees in technical and non-technical positions at a range of levels within each firm. I also interviewed a range of industry and culture *watchers* whose insights might offer important evidence or counter-evidence for hypothesis testing. In total, I conducted 100 interviews and gained over 130 hours of taped material. Each firm case study was further developed using secondary data sources (annual reports, memos, etc.) as part of a source triangulation strategy. In ten cases I was also invited to tour the firm and to observe general goings-on.

8   This case study sample is comprised of firms with the Standard Industrial Classification code 7371 (Computer Programming Services) because this subset of firms had the highest response rate (70 percent) in the initial survey.

9   This involved coding the data to break it down, recategorizing it, examining the links between groups, and then developing hypotheses with regard to the mechanisms and patterns that best explained the data. To make the analysis more robust, I also checked the credibility of my analyses with members of the group from which I originally obtained the data.

10 The settlement of the barren, harsh desert environment of the Salt Lake Valley necessitated a cooperative irrigation effort in an environment that would not have yielded to more individualistic efforts (Toth 1974).

11 Trust as "a psychological state comprising the intention to accept vulnerability based upon positive expectations of the intentions or behavior (of a partner)" (Hutt, Stafford, Walker, and Reingen 2000: 52).

12 Strategic partners defined in terms joint product development and/or R&D, or other self-identified formal alliances as outlined on firms' corporate websites or by my industry-respondents. Strategic partners for Mormon founded and managed firms compared with non-Mormon founded and managed firms: (i) survey sample: micro category: 4.1, cf. 7.0; medium category: 3.5, cf. 7.1; (medium-large category: 4.9, cf. 12.0); (ii) case study sample: 4.2, cf. 7.8.

13 Equivalent figure for the Mormon Intermediate firm is 49.1 hours per week. This pattern also remains when start-ups (firms less than 3 years old) are excluded from the analysis.

14 The holiday lengths calculated here are for software engineers below the Vice-President level who have been with a firm for three years.

15 US average: 11.7 days annual paid vacation for professional, technical, and related employees after three years service with the company, US Department of Labor (1996).

16 This examined over 300 establishments all of which employ over 20 staff.

17 Measured in terms of output while controlling for materials used, employee hours, age of equipment, industry, size, and employee turnover.

18 See also Radcliffe's concluding chapter, which argues that the role of culture in shaping creativity is one aspect that has been played down in development policy as it is difficult to plan for and encourage, but nevertheless has extensive social, economic and political impacts.

19 Nevertheless, I am aware of the debates surrounding the need for constructive confrontation in the firm, and whether this really *is* critical for innovation, given the success of Japanese firms based on very non-confrontational work cultures (see Ouchi 1981; Pascale and Athos 1982; Suzuki, Kim, and Bae 2002).

20 Nor by gender or sexual orientation.

21 In 1999 the LDS Church supported 58,593 LDS Missionaries in the field to 120 countries worldwide (Church of LDS/Deseret News 2000). Approximately 75 percent are young men aged 19 to 26. After eight weeks' training in Utah, Missionaries are sent out in pairs, on two-year assignments (18 months for females) to teach the LDS Gospel, win converts, and participate in community service.

22 Although, given that firms sometimes flout the requirements of Title VII of the Civil Rights Act (1964) in their deliberately hiring workers of a particular cultural background, the impact of legal frameworks is likely to be uneven across different firms.

23 The main methods outlined in this literature for culture change in the firm include: bringing in new management; developing new systems of rewards and punishments; implementing training programs; creating a new corporate image which employees must live up to; changing around people's positions within the firm; bringing in "new blood" through modified recruitment practices to introduce new modes of thought and alternative beliefs by which firms define and carry out their business; or through out-placement programs to immerse employees in fresh viewpoints, approaches, and attitudes.

24 Indeed, subsequent work on the role of masculinist corporate cultures among ICT firms in Cambridge's high-tech regional economy (Gray and James 2006) show how the same key individuals, strength-in-numbers, labor market, educational and legislative mechanisms are visible in the UK context.

9 Mobilizing culture for social justice and development: South Africa's *Amazwi Abesifazane* memory cloths program

Cheryl McEwan

Introduction

There has been a concerted effort by theorists in recent years to shift culture from the margins of development studies and to demonstrate that it is, and always has been, central to understandings of development processes and their impacts on peoples and societies around the world (Porter, Allen, and Thompson 1991; Allen 2000; Schech and Haggis 2000). Michael Watts (2003; also Watts, this volume), for example, argues that culture has always been at the center of post-war development theory. Colonial governmentality worked through local culture to gradually construct or attempt to construct a new sort of colonial subject and yet, paradoxically, traditional cultures were also seen as an impediment to progress, innovation and "development." Modernization theories of the 1950s and 1960s were unable to see beyond culture; theorists "read culture out of their own theory of the modern" (Watts 2003: 434). Cultural difference has been perceived to have consequences for growth, from the cultural particularities of the Southeast Asian "tiger" economies (Rigg 1997) to the perceived deficiencies of African moral economies (Ferguson 1994; Escobar 1995). More recently, as Watts argues, there has been a shift away from theories in which development describes a transition from tradition to modernity where cultures converge to those that see modernity and development as embedded within specific cultural contexts and are thus sensitive to the different starting points of the transition and its outcomes.

In other words, development is always culture and site specific – "it is irreducibly *cultural geographic*" (Watts 2003: 435). Development is now recognized as culturally and ethnographically grounded – in projects, development institutions and so on – and thus there is a need for an institutional ethnography of development. In addition, theorists and activists seeking to posit alternatives to dominant notions of "development" are sensitive to how alterity[1] and cultural diversity make a difference to imagining and reimagining development and in thinking about alternatives.

Despite this "cultural turn" within development, it might be argued that there is a large gap between how culture in the development process has been conceptualized and implemented. Rao and Walton (2004) acknowledge that cultural notions are now routinely incorporated into practice. However, they also point out that academics (they refer to anthropologists specifically), on the one hand, seem focused on critiquing development rather than engaging with it constructively while, on the other hand, policy economists either treat culture as emblematic of a tradition-bound constraint on development or ignore it completely. While the criticism that academics rarely engage constructively with development practice might be overstated, it would seem that there are some gaps between theory and practice that require closing. I also concur with Watts' argument that while a cultural geography critique of development is extremely powerful, the search for alternatives is

> often politically myopic, misreads the insights to be gained from radical or unconventional development theory . . ., has surprisingly little to say about economics and economic alternatives, and in some cases is simply politically reactionary and crudely anti-modern in ways that do no justice to the very idea of modernity itself.
>
> (Watts 2003: 435)

In this chapter I want to engage directly with these critiques of cultural approaches within development through a case study of a specific local development initiative in South Africa – the *Amazwi Abesifazane*[2] memory cloths project in KwaZulu-Natal.

I begin by focusing on the *Amazwi Abesifazane* case study to explore the possibilities of agency, identity, and memory in galvanizing culture through women's crafts and neighborhood groups – where culture is not necessarily fixed – and how this mobilization of culture is related to both economic and social justice. These connections point to the potential of a notion of culture-as-creativity as a possible policy avenue in new approaches to culture and development (see Chapter 10, this volume). In particular, I develop the notion that culture is an end in itself and a factor in the construction of value (Sen 1999, 2004), exploring both the ability of culture to inspire, express, and symbolize collective memory and identity and the ways in which cultural products contribute directly to well-being in more than an economic sense. The latter part of the chapter reflects on the significance of understanding the different trajectories for empowerment and social justice, reification of culture, and neo-liberal development that are put in train by these forms of organization at the local level. I also consider this case study within the broader conceptual framework of gender, culture, and development, highlighting the significance of approaches in development studies that put women at the center and culture on a par with political economy while retaining a focus on critical practices and movements for social justice (Bhavnani, Foran, and Kurian 2003: 2). I suggest that this framework is a positive move in closing the gap between theory and practice and, with its focus on social justice, in proposing potentially radical alternatives in reimagining development. The *Amazwi Abesifazane* example, I argue, demonstrates how local development initiatives based around gender, culture, and creativity might be effective both in galvanizing culture and promoting economic alternatives and social justice.

Gendered subjectivity, development, and social justice: the *Amazwi Abesifazane* memory cloths program

> Identifying women as vital, yet embattled elements within . . . cultural debate, assumes that true democratic process can only be realistically achieved if women are acknowledged as essential elements of the social and cultural discourse that must be economically affirmed and politically empowered. The Amazwi project has been conceptualized and implemented around the premise that women provide the consistent, essential social building blocks of the emerging South African society and provide stable, nurturing and emotionally affirming shelter, where the physical and emotional needs of children can be met.
>
> (Andries Botha, 2002, http://www.princeclausfund. nl/source_eng/gallery/botha.html; accessed 10 June 2004)

A significant aspect of inclusions and exclusions of citizenship in South Africa is the legacy of colonialism and apartheid and the multiple oppressions faced by marginalized groups. As South African sculptor and founder of *Amazwi Abesifazane*, Andries Botha, suggests, empowering women (and black[3] women specifically) is seen as critical to the successful transformation of South African society. The *Amazwi* project is based clearly on an essentialized and, therefore, potentially problematic notion of women providing both the bedrock of society and nurture and shelter for children. However, this is a strategic essentialism deriving from the fact that while race is no longer the principal line of exclusion defining relations between individuals and the state, the value systems upon which societies were structured during colonial and apartheid periods remain, to some extent, institutionalized. Black women remain most marginalized in a still profoundly polarized South Africa.

Fraser (1997) uses the phrase "bivalent collectivity" to refer to the economic and cultural forms of injustice and disadvantage that interlock, legitimize, and maintain each other. Despite being interrelated, "different forms of disadvantage have their own distinct logics and strategic responses" (Kabeer 2000: 86). Where disadvantage is largely economic, people are likely to mobilize around their interests and formulate demands in terms of redistribution. Where disadvantage is largely based on value systems, mobilization is more likely to be around questions of identity and demands formulated in terms of recognition. In South Africa, this produces a further tension: the logic of addressing economic disadvantage and of calls for redistribution is egalitarian, while the logic of addressing identity-based disadvantage and of demands for recognition is diversity. This tension is particularly problematic for those bivalent collectivities, like black women, disadvantaged by the interlocking dynamics of both resources and valuation. In the light of this, Young's (1990) notion of social justice has particular pertinence in South African citizenship, especially within feminist activism and gender advocacy groups. Social justice is understood in terms of freedom from oppression (constraints on self-development) and freedom from domination (constraints on self-determination) and enables a conceptualization of justice that refers to both redistribution and the development of individual capacities and group rights. Social justice requires not only the eradication of differences that construct relations of power, but the creation of institutions that promote and respect group differences without oppression. Significantly, a great deal of grassroots and community-based activism in South Africa is motivated by the pursuit of social justice and initiatives like *Amazwi Abesifazane*, as the quote above suggests, are clearly positioned within this context.

The *Amazwi Abesifazane* memory cloths project is part of community rehabilitation programs in South Africa aimed at promoting the healing and recovery of individuals and groups that have been affected by human rights violations, under the broader aims of the Truth and Reconciliation Commission (TRC). Although not directly linked to the TRC, the project claims this as its inspiration and source of influence. It is a unique project to provide black South African women from rural and urban areas with a vehicle to articulate traumatic experiences of the apartheid era, and to preserve and promote their creativity and memories. It was initially set up through women's organizations, including the Self-Employed Women's Union (SEWU) and the National Women's Coalition,[4] and other interested parties who facilitated workshops. A significant part of the workshop process was given to the importance of memory retrieval. Women from urban and rural areas of KwaZulu-Natal[5] who had experienced the trauma of the apartheid era were invited by trained coordinators from the same geographical areas to create a pictorial and verbal record of their experiences. Through the production of cloths, painful memories are transformed into creativity by embroidery, appliqué, and beadwork, drawing on indigenous arts and crafts, as well as Euro-American traditional sources, such as samplers and quilts (see Figure 9.1).[6]

Beadwork and the diverse material cultures of South Africa have a rich social history and the ways in which traditional skills are being used in memory cloths as part of an economic enterprise for poor people is of cultural and economic significance. Women's memory cloths share similarities with clay models made by Xhosa men in the Western Cape. They are "beguiling objects," designed to appeal as art but simultaneously encapsulating the history of a poor community; as such they need to be understood as framed within their historical processes and current social, cultural, and economic milieu (Morrow and Vokwana 2001). The use of beads is particularly significant because in South Africa (and especially in Ndebele and Zulu cultures) they are not merely for adornment but play a part in cultural rites (courtship, marriage, homage to ancestors) and modes of communication (status symbols, love-letters). The inclusion of beadwork within a formal archive is part of a growing recognition

Figure 9.1   An example of a memory cloth: no. 41 by Ntombi Agnes Mbatha. The accompanying text reads: "In 1992 my husband and I were going into town. Our son of 20 asked us to buy him some shoes. We bought him the shoes, but when we returned home our son was not there and then we heard he had been gunned down. We were so sad . . ."

of the place of ancient traditions and customs within contemporary nationhood. Sewing is also significant since, in many cultures and historical contexts, it has been used to communicate when, for various political and sociological reasons, oral disclosure has not been possible (this is discussed subsequently).[7] Initially, the intention of the project leaders was to assemble a national archive of 1,000 memory cloths to form a collective memory of life in South Africa up to the present day. Each cloth is an original work, to which is attached a profile of the artist and her story, written in her first language as well as English. According to Andries Botha:

> Through the creation of memory cloths, we are drawing on the collective experience of women who have known loss. Through the process of creation they will hopefully reach some level of catharsis through which they can grow both spiritually, emotionally and financially. This is a necessary, albeit humble, attempt to begin to transform the oral archive into a more formal record of South African history.
>
> (Cited in Anon. 2001: 7)

Although primarily aimed at memory retrieval, the project is also dedicated to improving the lives of women by encouraging peer support, nurturing dialogue at the local level, and developing women's self-employment industry that creates products to market internationally. There are, of course, some potential contradictions in pursuing both commemorative and market-oriented aims. Invoking memories of violence and trauma might not appeal to consumers, particularly those unfamiliar with the specific context of KwaZulu-Natal. It remains to be seen whether the desire to commemorate women's lives within South Africa is superceded by a desire to generate profits through international sales and by appealing to particular consumer groups. To date, the project has managed to do both successfully, but it will be interesting to see whether the nature of the cloths changes as the national archive is established and if the focus switches more towards market-oriented concerns.

The wider context of *Amazwi Abesifazane* is significant in bringing together culture and economy in a broadened notion of development, particularly as it pertains to impoverished South African women. It is part of Create Africa South, a non-profit, non-governmental organization established in 2000 in response to the diminishing funding base for arts and culture to promote and develop creativity in South Africa, particularly in historically disadvantaged areas (http://www.cas.org.za). With Andries Botha as President, Mazisi Kunene, one of Africa's senior poets, is Vice President. Principal funding is provided by the Prins Claus Fund in the Netherlands and working partners include African Art Centres in Durban and Pietermaritzburg, the Association for Women's Empowerment in Esikhawini, the Documentation Centre in Durban and the Self-Employed Women's Union. Creating a trust fund under the auspices of Create Africa South, but administered and contributed to by the participating women, is also an aim of the project. Profits generated by sales are intended to contribute to the trust fund; the aim is to use these profits to fund children's education (which is not free in South Africa – even the poorest parents have to pay a nominal amount to send their children to school) and thus provide wider benefits to livelihoods, particularly in rural areas. The project was launched at an exhibition of memory cloths held at Durban Art Gallery in June and July 2001, where as many cloths as possible were exhibited in an educational environment. The educational focus is on the process of memory retrieval, loss, women's issues (especially gender violence), the role of catharsis in healing a nation, and the role of women.

Once the national archive is completed, the intention is create a computer-based record of the original cloths and testimonies and to sell the entire archive. The proceeds of the sale will form the first major donation to the trust fund. The 2010 project goal of the trust fund is to acquire 5,000 stories and cloths and to create a permanent exhibition and educational program in an appropriate facility in KwaZulu-Natal. In addition to this, each memory cloth sold commercially through a networking system within the cross-cultural and international arts community will also contribute to the trust; to date this has included exhibitions at art galleries and museums in South Africa since 2001, exhibitions in Nashville, Chicago, and Waltham, Massachusetts during 2003 and 2004, and in Lisbon, Portugal in May 2004. After initial surprise and some bewilderment that their cloths could be considered art, the participating women articulate pride and pleasure at the fact that their cloths are both displayed and purchased within South Africa and beyond. They also attach considerable significance to the fact that their experiences are being recorded and their stories heard.[8] The purpose of the trust is to promote similar self-empowerment projects within localities and especially for individual mothers experiencing difficulty in educating their children, enabling women to develop their own creative, economic initiatives that will lead to personal autonomy.

Documenting women's lived experiences

The first cloth made by each woman forms part of the 5,000 cloths being collected for the national archive. At present, the program is based in KwaZulu-Natal but the intention is that it will be extended into other provinces. The women were invited to contribute a single memory that they felt was important to the project, and the nature of these memories is significant in documenting women's experiences of both apartheid and the transition to democracy. Perhaps not surprisingly, many cloths reveal relatively recent memories of violence in KwaZulu-Natal. Between the release of Nelson Mandela in 1990 and the first democratic election in 1994, the National Party used its command of the state machinery in a merciless fashion to fuel effectively a civil war in the province. As Asmal argues:

> Hit squads intensified, attacks on innocent commuters were carried out by the Civil Cooperation Bureau (an Orwellian name for the state security body) and, in the killing fields of Natal, the regime funded and

fuelled a civil war which it presented to the world as "black-on-black" violence. As many people were lost to political violence between 1990 and 1994 as were killed during the entire previous history of apartheid, dating back to 1948.

(Asmal 2000: 9–10)

The TRC hearings ultimately failed to force an acknowledgement of the central role that the state machinery played in the violence between rival ANC and IFP factions. However, the memory cloths project is ensuring that the voices of those whose lives have been irrevocably scarred by the violence are embodied within the national archive. It is not surprising given the level of violence in KwaZulu-Natal that memories of this time are represented frequently in the cloths. The bewilderment and trauma of these events are encapsulated in the written accounts that accompany the pictorial representations:

The day I will never forget is when there was violence between two organisations, Inkatha and the ANC in 1985. On June 16 people were ransacking the shops . . . It was an extremely agonising experience for me when they took my cousin Khayini and put a tyre round his neck.

(Sindisiwe Hlongwane, #266)[9]

As a result of friction between two political organisations violence erupted on 25th December 1995. This happened on the South Coast of KwaZulu-Natal, at Shobashobane. A certain kraal was attacked by gunmen and the head killed . . . The bodies of the deceased were doused with petrol and set alight. The house was also burned and the children left destitute.

(Christobel Ngcobo, #295)

What a life of unhappiness we lived in 1990, seeing children being shot and necklaced.

(S'bongile Makhanya, #133)

At Mkhomazi, in the area of Zembane, two political organisations, the ANC and IFP were fighting. It was on a Sunday morning when IFP members attacked us. They shot and killed my four bothers . . . We fled the area and took a taxi to seek refuge at KwaMashu K section where we now live. We then decided to go to Durban to start selling fruit in Victoria Street trying to fend for the orphaned children.

(Zendile Nzama, #222)

Violence broke out in 1991 between IFP and the ANC at KwaMaphumulo. Many people died there including my uncle . . .

He left behind his wife Mama Ndlovu and two children . . . MaNdlovu
gets money by ploughing mealies fields for other women in the area
and fetching water from the river for them. It's sad that a woman
suffers this much just because of politics.

(Mavis Ngcubo, #657)

These lived experiences of violence and trauma within families
and the dislocation that often resulted were largely absent at the TRC
hearings. Unlike the grand narratives of apartheid, the memory cloths
reveal that everyday survival struggles and violence against women
are often foremost in the memories of black women.

Many of the cloths relate less directly to the injustices of apartheid
(although several record the traumas of families being dispossessed
of land and forced removals) but to the daily struggles for survival.
In particular, natural disasters and their impacts are recounted. For
example, Celani Nojiyeza's cloth records one of her most vivid
memories: "In 1982, the sun was so hot that it burnt all my mealies
and there was no food for my family. At that time, I was dependent
on the food I grew to feed my family, so it was a disaster" (Celani
Nojiyeza, #424). Similarly, Phille Mabaso describes the after-effects
of the 1987 floods in Pietermaritzburg:

People who suffered most after the rains were the women who had
to look after the orphaned children . . . Women had to shoulder the
responsibility of dealing with property damaged by water, falling trees,
stones, at the same time giving love to their children . . .

(Phille Mabaso, #824)

Significantly, nearly all the cloths feature homes, reinforcing the
significance of women's roles in domestic and community life. The
loss of homes is also a persistent theme within the project, both as
a consequence of apartheid and of factional violence during the late
1980s and early 1990s. For example, one woman describes the effect
of her family being evicted from their land by white farmers in 1970:
"Even today I still have a loathsome feeling for white farmers for the
inhumane treatment they inflicted on my family" (Tuleleni Tenza,
#216). Another woman describes her memory cloth as follows:

This is the incident that took place at Mkhomazi, in the area of Qiko
on the 20th September 1987 . . . when a faction fight . . . broke out
between the Matshobho and Mkhono factions. I lived with the Mkhono
faction. People from Matshobho faction came and shot my uncle . . .
in the chest and he died instantly. I, together with my children . . . fled

the area. We took a taxi and went to Durban. We now live in
KwaMashu township.

(Nokuthala Khomo, #207)

Men's violence against women is also a common theme:

It was on Friday of November 1987 when my grandmother
MaMhlongo, told me that I, together with my children, were going to
be killed on that night . . . The reason for us to be killed was to punish
my husband . . ., as a retribution for the alleged killing of my elder
uncle in law . . . We left our home never to return. I was very
disturbed because I left behind our livestock, furniture and all the
clothes except for what I carried in a small bag.

(Sibongile Myeza, #211)

My mother was building a house. They had been paying the builder all
the way through the building of it and when the house was finished the
builder demanded more money. The money was now finished and an
argument ensued. The builder became very angry and shot my mother
dead.

(Pumazile Khubisa, #356)

Perhaps the most poignant story of the violence perpetrated against
women is that of Nonhlanhla Hnyandu, who was killed just two
months before the opening of the Durban installation in April 2001.
She had created a memory cloth for the *Amazwi* archive in the first
workshop held at Richmond Farm/Ntuzume B. She also co-ordinated
a subsequent workshop at Phindangane in Richmond Farm and was
seeking to raise money to educate her children. A simple memorial
in the installation at Durban Art Gallery explained the circumstances
of her death:

Nonlanhla was a victim of domestic violence. Her boyfriend stabbed
her to death with a screwdriver in front of her three children.

Despite cultural silences around sexual violence, several women
use the memory cloths to relate their experiences. One woman
recounts:

I will never forget the day I was raped by a stranger . . . I was so angry
and depressed because he had taken away what I had treasured most –
my virginity . . . When women's organizations were formed, we hoped
the incidences of rape will be little, but it's not getting better, instead
it's worse than before.

(Mathombi Nxumalo, #828)

Unsurprisingly, the violence and devastation associated with HIV/AIDS, which is widespread in the rural areas of KwaZulu-Natal, is also a prominent theme of the exhibition. For example, Makhosi Khanyil tells the story of Gugu Dlamini, aged 22, who was open about her HIV-positive status, even going on TV to discuss her illness:

> One day as she was coming from Durban, and soon after alighting from the train at KwaMashu station, she was stoned and stabbed to death. Nobody came to her rescue because she was HIV+.
> (Makhosi Khanyil, #798)

The memory cloth (Figure 9.2) illustrates her bloody murder. Other testimonies reveal the violence and ostracism faced by women with HIV/AIDS:

> On June 12, 1995, my uncle Zolani found out that he was HIV+ ...
> My uncle decided to blame his wife Nandipha even though he was sleeping with several women. He came home drunk one day, and with his friend Vukan, he took his wife to the bush where they stoned her until they thought she was dead. Nandipha somehow managed to stagger to a nearby house ... She died on the way to hospital.
> (Buselani Nene, #838)

> Nozuzo Gumede was my best friend who lived at Richmond Farm. In 1997 she died of AIDS. She was 19 years old. She had nobody to take care of her during the day because her mother was at work. Her mother asked me to help my friend. Nonzuzo was so weak that she slept all day and was unable to do anything. I used to wash her, spoon-fed her and change her clothes as well as her linen.
> (Zama Zulu, #842)

The text accompanying another cloth states simply: "The whole family died of AIDS" (Thembisile Shozi, #357).

Although the vast majority of cloths relate memories of dispossession, loss, trauma, violence, and death, several also record happier memories about everyday rural and community life. Others celebrate the ending of apartheid and the promise of happier times. For example, Margaret Thakane Lesoma produced a cloth entitled "Chains are Broken." Her testimony states:

> Previously we were denied our rights but today all that has changed. The time has come for us to enjoy, discover and explore the true

Figure 9.2   Makhosi Khanyil, no. 798.

meaning of so-called life. The day has come when all prison doors
open and we are free from prison chains. Freedom! Freedom! How
long we've been waiting for you.

(Margaret Thakane Lesoma, #25)

Other cloths celebrate the strength of women. For example, Busi
Mlotshwa's cloth, entitled "Portrait of a Zulu Woman," is
accompanied by the following testimony:

She has that inner beauty that only those close to her can see and
appreciate. Those lips may not have a smile, but they have put smiles
in many faces. She has acquired wisdom in all that she has witnessed
through those bold beautiful eyes. Her family is always surrounded
with her warmth, security and understanding. She is determined,
full of life.

(Busi Mlotshwa, #315)

It would seem that the *Amazwi* project is playing a significant role in the construction of value and in demonstrating how culture might be mobilized to inspire, express, and symbolize collective memory and identity. The project also illustrates the ways in which cultural products contribute directly to well-being in more than an economic sense. The next section discusses this in more detail.

Memory, development, and social justice

Through its enabling of the voices of marginalized women, its restoration of their historical agency to the national memory-archive, and its emphasis on catharsis, economic empowerment, and social transformation, *Amazwi Abesifazane* is playing a small but significant role in continuing the truth and reconciliation process, both at the local level and in terms of the national archive. The reconstruction of post-apartheid South Africa and the development of a viable democracy requires acknowledgement of the central role that women play in consolidating the building of nation, homes, and neighborhoods. Memory cloths are a powerful way of acknowledging the agency of ordinary women. *Amazwi Abesifazane* pays tribute to the sacrifices made by women that made democracy possible and records lived experiences of daily struggle both in the past and in the present. Its radical potential lies in the fact that it can be used to counter the erasure of women's historical agency in addition to the erasure from dominant discourses of the polarizing effects of neo-liberalism and the devastation of HIV/AIDS. It draws on the rich historical tradition of crafted creativity as a means of communication that is at once intimate and public. Small workshops allow the participants to share intensely personal events and emotions, which are then used to shape the individual cloths. The intimate disclosure of traumatic experience, which proved impossible for women through the TRC hearings, sets up a necessary catharsis that enables women to speak publicly about their experiences and acknowledges the importance of memory as part of the oral archive of South Africa. The abstract encoded language evidenced in indigenous beaded craft techniques of African women helps facilitate disclosure of events and emotions that might otherwise be difficult to articulate. The project, therefore, is an attempt to engage an oral archive of women's experiences that remains undisclosed for the same reasons that the TRC remained largely silent on gender violence. It is clearly given impetus by the post-apartheid legislative exercise of reconstruction

and memory as a mechanism of reconciliation, but also by the need
to countervail the gender processes that continue to both erase
the voices and agency of historically disadvantaged women from
national memory-archives and position black women at the margins
of economic and cultural processes.

The memory cloths program is one of several examples of archival
and crafts-based developmental projects in South Africa and
complementary or alternative strategies to the truth and reconciliation
process, including religious and non-governmental organizations that
are actively involved in creating, documenting, and preserving
collective memories at the local level as well as pursuing economic
and social justice (see, for example, Chirwa 1997, Walder 2000).
Radical memory projects, rooted in both the discursive and material
empowerment of previously oppressed peoples, are significant in
terms of how they relate to broader economic processes and how
they might inform approaches within development studies. Regarding
economic processes in South Africa, memory projects are particularly
important given the pressures by vested interest groups to produce
what Simon Lewis (2000) calls "sanitized" versions of history
to buttress economic policy. Lewis argues that contemporary
representations of South Africa and the apartheid past are very often
sanitized, primarily by the business community, for international
consumption and in this sense cultural change has not occurred. He
discusses, for example, the renewed confidence in the significance of
sport, expenditure of money and organized leisure, the recycling
of old images of wild animals and exotic landscapes, and the
transforming of unacceptable racial exclusivity into acceptable class
exclusivity. In contrast, cultural workers such as policy-makers,
historians, teachers, writers, and artists see the local processes of
remaking and re-presenting South Africa as "ongoing projects
of considerable urgency, complexity, and precariousness, projects in
which the often uncomfortable de-sanitizing of apartheid era stories
and images plays a key role" (Lewis 2000: 47). This involves
acknowledging and confronting the violence of apartheid through
education, popular culture, radical memory projects, and art, theatre,
and writing projects with historically disadvantaged groups such as
black women and youths.

The gap between "entrepreneurial empire-builders" and "community-
minded nation-builders" has widened since 1994. As Lewis attests,
while historians and artists, and state-supported processes like the
TRC, have been laboring to create a "new" South Africa that can

come to terms with its violent past and the suppressions and misrepresentations of that past, "South African business has tended to want to take 1994 as a marker of an end of history or an end of politics" (Lewis 2000: 47). These divisions are potentially dangerous, especially in a context where a racialized polarization between rich and poor still persists. However, the different visions of development need not be incompatible and there are examples where white-owned businesses have acknowledged the violence of the apartheid economic system, embraced notions of "black economic empowerment" and are working in partnership with worker's groups to build capacity.[10] The *Amazwi* project demonstrates that commemorating the past and working toward social justice in the present need not be counter-productive in an economic sense and can play a role in improving livelihoods, at least at the local level. Indeed, acknowledging the significance of economic and social justice requires that the memories of the people who experienced the "minutiae of social and community life under apartheid" should not be lost, "erased in old or new forms of grandiose history writing" (Field 2001: 119) or forgotten by business. As Andre Brink (1998) argues, memory is a means of excavating silence; it also provides the first step in moving towards economic and social justice.

The *Amazwi Abesifazane* case study also highlights the potential of approaches in development studies that, as Bhavnani, Foran, and Kurian (2003) have called for, put women at the center and culture on a par with political economy while retaining a focus on critical practices to bring about social and economic justice. For example, they demonstrate the importance in developmental terms of creating postcolonial, post-apartheid national memory-archives. First, in memorializing the past, it is important to keep multiple versions of history alive and "not to privilege, as has so often been done, a few master narratives that offer sense of unity at the cost of ignoring the fracture and dissonance" (Nuttall and Coetzee, 1998: 14). Resisting various kinds of amnesia is essential to the creation of a shared past, a shared sense of national and communal belonging, and the prospects of agency in social and economic development. As Lewis (2000: 47–8) argues, amnesia allows South Africa's business leaders to maintain their wealth and privilege while urging "all South Africans to bury the past unconditionally." This amnesia threatens to erase past economic exploitation "even as the actual bodies of apartheid violence are being exhumed" (Lewis 2000: 48). Projects such as *Amazwi Abesifazane* play an important activist role in countervailing attempts to sanitize accounts of the past and blithely to sell South Africa abroad "as if it were fully rehabilitated already" (Lewis 2000: 48). In this sense, while the project could be thought of as generating culture-as-product (Chapter 10, this volume), where cultural difference illustrated by handicrafts and folklore generates income through international sales, it is arguably more productively thought of as culture-as-creativity, where culture is perceived as a resource by the participants, a flexible combination of tradition and modernity that does not rely on a problematic nostalgia for the past or a presumed purity of non-Western cultures. The project engages with efforts to mobilize poor women's concepts of culture and social change and thus has the potential to provide an alternative viewpoint on development rather than work within existing (neo-liberal) frameworks as culture-as-product. In its efforts to fill the "great blank spaces of still repressed memory in South Africa" (Lewis 2000: 55) and to bring about social and economic justice it also raises the possibility of more radical and less problematic notion of the role of culture in development.

Second, projects like *Amazwi Abesifazane* facilitate the preservation of collective memories of traumatic events and the possibilities of healing in the process of development, especially in post-liberation or post-conflict contexts. As Duvenage (1999: 17) attests, ". . . the identity of the present and future South Africa will not escape a moral obligation to the past"; constructing and preserving truly collective memory is, therefore, of the utmost importance. Collective memories, of course, have certain shortcomings. They raise questions about how collective they are and whose memories they are. There are also dilemmas about whether the memories of victims take precedence over those of perpetrators, or if myths of innocence and victimhood construct powerful obstacles in the way of confronting unwelcome facts. Chirwa (1997) argues, however, that because collective memory goes beyond an individual account, and has historical and emotional relevance, connecting seemingly discrete events in a cause-and-effect manner, it becomes part of the process of healing, reconciliation, and reconstruction at both the individual and communal levels. Therefore, Ignatieff (1996) is correct in arguing that nations do not have consciences, identities, and memories like individuals; thus, they cannot be reconciled to their pasts by replacing myths with facts as can individuals, nor can they be healed by working through traumatic events or memories. However, collective memory through the creation of national memory-archives can be an effective tool for reconciliation and healing for individuals and groups. As Stanley (2001) argues, newly emerging truths have had impacts at personal levels (individuals feel for the first time that they have been listened to by the state) and at collective levels (stories have challenged the traditional perceptions of groups and individuals). For black women, the simple act of publicly telling a story in their own language has provided, and continues to provide, a sense of symbolic liberation (Bozzoli 1998).

Third, locally based projects are essential in ensuring that women's voices can be incorporated into national projects of remembering and notions of belonging and their agency prioritized in development processes. It is particularly important to consider the role of gender in erasing the historical agency of women and denying them agency in the present. Without spaces for the articulation of memory, black women's citizenship, in terms of social standing and belonging, continues to be compromised. The role of women's personal testimony and agency in shaping the nation and citizenship is particularly important in a country such as South Africa, where the legacies of colonialism and apartheid have effectively silenced black women's voices. Those legacies persist today and the relating of personal testimony by black women is extremely important to their being able to claim space in the imaginings of the nation, both historically and as citizens in the present. As socio-historical processes, intimate disclosure projects ultimately become part of broader social movements that shape the dominant culture. Moreover, the participants produce valuable public discourses. The projects are potentially moments in the creation of a radical discourse, revealing a great deal about black women's experiences of racial and gender oppression and the ways in which their citizenship is constantly mediated through gender norms and power relations. As Graybill (2001: 6) argues, there is a need to "pierce that which destroyed or constrained women's voices." It is imperative that women do speak out, and only when they do will they begin the healing process. Women's private suffering needs to be made "visible as social suffering, enabling them to stake their historical claims and thereby restore their dignity" (Ramphele 1996: 114); this is clearly critical in building capacity and recognizing the centrality of women's agency within grassroots development.

Individuals, groups, and institutions have collaborated in mapping the past and recording personal histories in the light of the findings of the TRC and criticisms of the process (de Kok 1998). The challenge for grassroots memory projects is arguably twofold. First, the absence of black women in the histories of apartheid needs to be continually addressed since it has a direct bearing on their role in the still unfolding transformation of South African polity and society. Second, memory projects must also be connected to notions of social justice in order to be truly transformative. In developmental terms, decisions for societal transformation cannot ignore the gendered experience of conflict and violence. To prevent further repression and discrimination, there has to be a form of social justice. The government's commitment to human rights requires it to promote women's equality of opportunity, economic security, and protection from violence. The challenge is in allowing those people who do not have access to state-level processes such as the TRC to create collective memories to facilitate the process of healing, reconciliation, and reconstruction. The localization of the truth and reconciliation process and its associated grassroots memory projects such as *Amazwi Abesifazane* plays a critical role in this. Creating "postcolonial" archives, where previously silenced and marginalized voices are brought to the fore in an effort to counter the epistemic and material violence of erasures under colonialism and apartheid is clearly important. It plays a significant role in constructing individual and collective identity given the inability to conserve memory within distressed communities that often results in the erosion of local knowledges. It also points to the significance of culture in questions of gender and development and opens up the possibility of new ways of imagining development, which I now discuss further.

Closing the gap: culture and development

Projects like *Amazwi Abesifazane* offer new perspectives on locally based initiatives concerning empowerment and social justice, drawing on the centrality of culture to development and proposing an alternative to neo-liberal development with an emphasis on grassroots rehabilitation. They relate to and inform broader notions of gender, culture, and development, while focusing on critical practices, pedagogies, and movements for social justice (Bhavnani, Foran, and Kurian 2003: 2). Within this approach, culture is neither fixed nor reified. Instead, it draws on a notion of culture as lived experience; not simply as a set of habits or traditions, but as a way to understand how people actually live their lives.

This has significance for how we might conceive of development, since it deploys a notion of culture as conceptualized by Raymond Williams – as a "structure of feelings." It employs an agentic[11] notion of human beings, based on dynamic sets of relationships through which inequalities are created and challenged, rather than a single notion that resides within an individual, group or nation (Williams 1960; Hall, Hobson, Lowe, and Willis 1990). Conceptualizing the relationship between development and culture in this way avoids many of the problems in much of the gender and development literature that either glosses over the significance of culture or sees women as victims of culture. As Prabhu (2003: 242) argues, Raymond Williams' notion of culture brings with it debates about how other terms such as economic, material, gender, class, race, or ethnicity, for example, articulate themselves within this definition. She suggests that any understanding of agency or resistance within particular cultures must necessarily relate to how exactly this notion of culture is defined. The *Amazwi* project demonstrates the significance of understanding how people live and experience structures of relations and the importance of emphasizing creative and historical agency.

The *Amazwi Abesifazane* case study also supports Prabhu's contention that a more culturally informed approach to gender and development is required because of the urgency to move away from economistic analyses, specifically in the context of Third World women, in order to understand women's lives and agency in all their complexity (Prabhu 2003: 242). It usefully points to a notion of culture as "enunciation" rather than as epistemology, where the "enunciative is a more dialogic process that attempts to track displacements and realignments that are the effects of cultural antagonisms and articulations – subverting the rationale of the hegemonic moment and relocating alternative, hybrid sites of cultural negotiation" (Bhabha 1992: 443). Memory projects are a process of generating culture that is anchored in specific social contexts, as the specific memories recorded by the *Amazwi* participants demonstrate. However, they are also tied closely to disruptions of static ideas of culture and tradition. They work in contrast, for example, to cultural stereotypes about South Africa, such as those invoking exotic rural landscapes, game parks or the ubiquitous "Zulu warrior," that are often promoted by businesses and tourist agencies. As Prabhu (2003) argues, the consideration of these disruptions that are enabled by oppositions and ongoing practices within the circumscribed space upsets an understanding of culture as a given that pre-exists action.

Such an understanding provides a means for questioning purely economistic conceptualizations and opens up possibilities for reading development that can take into account multiple spaces, especially those inhabited and generated by women, and unconventional types of action, with a new understanding of "production." This raises the possibility of a more complete, nuanced understanding of women in the context of development, which engages their creative resistances to various hegemonic forces. As Appadurai (2004) argues there is a need to strengthen the capacity of the poor to exercise "voice" – treating voice as a cultural capacity – because it is not just simply a means of inculcating democratic norms but of engaging in social, political, and economic issues in terms of metaphor, rhetoric, organization, and public performance that work best in their cultural worlds. Appadurai argues that the cultural contexts in which different groups live form the framework of the "capacity to aspire," which is not evenly distributed. Voice and the capacity to aspire are reciprocally linked and are significant to issues of development. For the women of *Amazwi Abesifazane*, having a voice in creating stories of the past and accounts of the present informs their aspirations both to claim meaningful citizenship and a sense of belonging in post-apartheid South Africa and to improve the material circumstances of their lives. By aspiring to improve their lives in this way they are also, in turn, asserting their cultural capacity. This raises the importance of closing the gap between cultural analyses that too often have eschewed the significance of the economic, and more reductive economic approaches that have failed to consider lived experiences and notions of social justice. For Prabhu (2003: 253) this involves not only a "cultural shift in analyses relating to areas already considered in development studies," but rather, would involve "parallel enlargement of the field of inquiry itself, which would require legitimization of other bases of information on, and insight regarding, its central question."

Conclusions

In this chapter, I have attempted to demonstrate the significance of culture in development through a specific case study in South Africa. I want now to make a few brief points by way of conclusion. First, the *Amazwi Abesifazane* example demonstrates that development projects that bring together cultural and economic approaches have radical transformative potential since they are reflective of the strong commitment among grassroots activists and artists to create a "rich knowledge based both on women's experience, and on the complexity of gender relations as relations of production/reproduction" (Bennett 2000: 5). These knowledges involve challenging the critical under-representation of women (in histories and elsewhere) by giving rigorous attention to their complex and multiple experiences. They also challenge the androcentrism of African knowledge production by asserting that African women can be producers of knowledge. This requires a commitment to applying knowledge to create social justice. Memory cloth projects are part of those knowledges that can be activated to bring about gender justice. They reveal the persistent economic and political discrimination against women and the intransigence with which unequal gender roles are inscribed into everyday life and socio-legal practice to compromise women's citizenship; they also work to counter these discriminations, revealing the various ways in which South African women are attempting to resist this to claim material and metaphorical spaces of citizenship, a sense of belonging, and a means to improve their livelihoods.

Second, apartheid era human rights abuses, and the willing participation of global economies in prolonging the legacies of apartheid inequities and injustices, emphasize the importance of the flawed, fragile role of memory and oral disclosure, of constructing a postcolonial archive and of the possibilities of culture-as-creativity. The *Amazwi* cloths demonstrate that, to a large extent, the lived experiences of those who bore the brunt of state brutality have not altered significantly, despite the promises of democracy. In allocating a space for those who have been previously silenced, the project is playing a small but significant part in affirming that individual experiences of trauma are important at a societal level, both in terms of the past and the present. This "disclosed" oral history informs the legislative framework of amnesty and forgiveness required by the South African Constitution and the importance it places on the culture of human values. Documenting such oral experiences by means of an archival, legislative, and creative process is essential to acknowledging the complexity of democracy in South Africa, and of women's role as active citizens. Forms of public representation by women are potentially a politically vigorous means of constructing visibility and accountability. As Bennett argues, echoing Zeleza, in order "'to imagine a South Africa free of gender injustice, and to understand the intersecting vectors of racism and misogyny' in the present, 'access to women's experiences of the past is critical'" (Bennett 2000: 46). Attempts, like the *Amazwi* project, to construct gender-sensitive archives that also have a positive impact on grassroots development will continue to play a significant part in this.

Finally, in broader terms the chapter has demonstrated that the gap between how culture in the development process has been conceptualized and implemented might be closed by a greater understanding of grassroots development and reconciliation programs like *Amazwi Abesifazane*. Furthermore, this case study has attempted to answer the various criticisms (by the likes of Jackson 1997; Watts 2003) that cultural approaches in development have had little to say about economics and economic alternatives. As Rao and Walton argue:

> In recognizing the role of culture as fundamentally dynamic, endogenous, changeable, both forward and backward looking, and affecting both the ends and means of development, we see an acceptance of the social embeddedness of economic action and the economic embeddedness of social action . . . [C]ulture affects power relations within a society and is therefore fundamentally linked with the perpetuation of inequality.
>
> (Rao and Walton 2004: 30)

This is, of course, a small case study in a very specific developmental and transformative context. However, it raises a number of significant issues around the complexities of culture, development, gender, and social justice. Watts (2003) argues that a critical cultural geography of development should be capable of understanding actual development practices in cultural (semiotic, representational, discursive) and spatial (regional, territorial, global) terms. Although there is still much to be done in this regard, the *Amazwi* case study demonstrates the potential of galvanizing culture in development projects and how this mobilization of culture might be intrinsically connected to economic development, issues of agency and empowerment and the pursuit of social justice.

Acknowledgments

The research upon which this paper is based was funded by ESRC (R000223286) and was conducted in May and June 2001, while I was a visiting researcher at the Centre of Industrial Organisation and

Labour Studies/Sociology, University of Natal, Durban. I owe a debt
of thanks to Debby Bonnin and Richard Ballard for their friendship,
hospitality, and encouragement and to the women of *Amazwi
Abesifazane*. I am also grateful to Sarah Radcliffe for her close
reading of this chapter's first draft, and detailed and helpful
comments (the usual disclaimers apply).

Notes

1   By alterity, we mean "otherness," referring to the systematized construction
    of difference between groups of people on the basis of such factors as "race,"
    ethnicity, or culture. Alterity has been central to development theory and
    practice, with the construction of First World/Western/Northern/advanced/
    developed/modern "self" and Third World/Southern/backward/developing/
    traditional "other" framing interventions in both colonial and postcolonial
    contexts (see Power 2003).

2   This translates from isiZulu as "Voices of Women." The discussion here
    is drawn from information garnered through interviews with members of
    Amazwi Abesifazane at Durban Art Gallery (June 23 and 24, 2001), from
    the organization's website at http://www.voices.org.za and that of its supporting
    NGO, Create Africa South, at http://www.cas.org.za.

3   I use the term "black" as inclusive of all women of color, while being mindful
    of the sensitivities associated with such terminology.

4   SEWU was a new trade union to support self-employed women in South
    Africa. It was established because the interests of self-employed women,
    especially those employed in the informal economy (as many women are in
    South Africa), were not addressed within the traditional trade unions. The
    Women's National Coalition was launched in 1992 to bring together women
    across the political, economic, social, religious, and cultural spectrum and
    to identify, through a research program, the changes necessary for their
    emancipation in the post-apartheid context. Both organizations were established
    at a time of intense mobilization by diverse women's groups in response to the
    ending of apartheid and the recognition of the need to assert women's agency
    and presence in the "new" South Africa.

5   KwaZulu-Natal is one of the nine provinces of South Africa, incorporating the
    administrative area of Durban (eThekwini) and the Zulu Kingdom.

6   There are also similarities with the tapestries, known as *arpilleras*, made by
    women in Latin America. These originated in Chile in 1974 to commemorate
    women's experiences of the violence of the Pinochet regime and to register
    protest, particularly at the "disappearance" of relatives. The art form spread
    to Peru, Bolivia, Columbia, and Ecuador and is now produced primarily to
    generate money, particularly through overseas sales and sales to tourists (see
    Agosín 1987, 1996; Gianturco and Tuttle 2000).

7   See the discussion of the secret messages and symbols stitched into Red Cross
    quilts by British women POWs in Japanese camps during World War II in

Archer (1997). There are also some interesting parallels with Aborigine women's secret knowledges (Jacobs 1996).

8   Informal author's interviews with women of *Amazwi Abesifazane*, Durban Art Gallery, July 16, 2001.

9   All references to the cloths relate to the *Amazwi Abesifazane* Memory Cloths Exhibition, Durban Art Gallery, June 2001.

10 The issue of black economic empowerment is deeply contentious in South Africa (Black 2002; Heese 2003; Iheduru 2004). However, initiatives like Rooibos Tea at Wupperthal and Thandi Wines (both in the Western Cape) have made strides in involving black workers in the production process, in building capacity in terms of business knowledge, and in profit sharing. There has certainly been no wholesale transfer of ownership of the means of production, but efforts are being made by some white-owned businesses to transfer some land, resources, knowledge, and profits to historically disadvantaged groups.

11 Agentic infers that individuals have agency, or the capacity to exercise control over the nature and quality of their lives.

10 Conclusions: the future of culture and development

Sarah A. Radcliffe

Development has experienced a cultural turn and its language, paradigms, and actors are to a profound degree now engaged in understanding the cultural field within which development occurs. The reasons for this are varied, depending on diverse actors' standpoints – African writers are concerned about the loss of historical-geographically specific ways of organizing society and meaning, while policy-makers offer cultural alternatives to failures of the past. The decentering of the West through postcolonial and feminist criticism combined with the experiences of East Asian Tiger economies resulted in a revaluation of development's implicit values and a questioning of the Westward trajectory that undergirded much development practice. Yet culture as a factor remains wide-ranging; although anthropologists and cognate disciplines find it hard to pin down, they are agreed that culture is embedded in economies, politics, and racial formations. As material culture and structures of feeling, as well as specific forms of (re-)production, culture inevitably informs economic (and reproductive) life although it remains highly contested and becomes entangled in racial hierarchies and patterns of exclusion.

Development's cultural turn – as pointed out by a number of contributors to this volume – has occurred critically in the context of neo-liberal policies which have gained a global hegemony, albeit with local and national variants and resistances. After the harsh structural adjustment policies of the 1980s that marked the first phase of a neo-liberal project, the cultural turn in development marks a further instantiation of the "post-Washington Consensus" through which neo-liberal patterns of governance, production, and social reproduction are deepened and extended across new spheres of society and space. Yet for all its newness, cultural development draws on the meanings and powers invested in concepts that have a much longer history, namely community and scarcity, as Michael Watts describes in his chapter. Culture is often equated with a community; a community has culture, just as culture rests in a community. Examples of cultural development from Ghana, Ethiopia, and the Andes, described in this collection's chapters, illustrate how consistently these equations between culture and community are made in policy and development practice (even when the sums don't add up). Likewise, Watts shows, the Malthusian specter of scarcity has worked its way down through economics and successive generations of policy to frame discussions about the conduct of economic actors and their interrelationships. For centuries, this framework excluded discussion of social subjects in their non-monetary interactions, their voluntary, reproductive, and ritual activities, and in meaning-laden universes. Postcolonial critiques of economics' normativity and post-development's rejection of metaphors of scarcity can potentially lead us out of the articulation of conduct-biopower-money into more empowering and creative livelihoods. Yet, as Watts argues in his chapter, the ongoing production of neo-liberalism's hegemony reveals a formidable arsenal that would need to be challenged and engaged.

Development practice, of course, operates within the parameters set by the uneven expressions of capitalist political economy, where culture too has a new prominence. In the switch from a production-oriented to consumption-oriented economy, culture gains a heightened salience in differentiating consumers and driving the ever-changing goods through which modernity and identity are appraised (Harvey 1989; Jameson 1991; Comaroff and Comaroff 2000). In the mid-twentieth century, countries' cultures provided the infrastructure for belonging, being the material expression of a supposed national community, and the basis for mobilization for development goals. Culture and development – however intricately interrelated – were kept at arm's length. Now by contrast culture has gained the status of being an additional value – given to a place, a group, an institution, or a piece of material production. According to the UNESCO Decade on Culture and Development, cultural pluralism is "one of a country's most fundamental and creative aspects" (citied in Davis 1999).

Development's cultural turn, as discussed in Chapters 1 and 2, has – over the past decade or so – had an impact on development policy and practice on the ground with a number of different initiatives and interventions being informed by actors and paradigms which treat culture seriously as a factor. Through case studies and thematic discussions, the chapters explore in detail how development policies are being negotiated at a variety of scales by different actors and institutions, and how culture is being interpreted and mobilized. Working on the basis that culture has become one component in development's toolbox, the chapters analyzed how this general commitment to bringing culture into development is played out on the ground in highly diverse settings, and in the context of struggles over development *and* culture. In other words, the chapters describe how cultural development is put into practice in specific programs in particular locations, the why, how, when, and where of culture and development.

Cultural development has been used to inform community-based natural resource management programs. Whereas the second phase of Ecuador's Indigenous Development project has only just introduced a NRM component, Ethiopia's development practitioners have been talking about and using "culture" for a number of years, as described by Elizabeth Watson in her chapter. While the "traditional" institutions of the Borana group in this case appear to offer an ideal cultural tool for development practice – in that they are assumed to be legitimate locally, enduring, non-politicized, and culturally appropriate – in effect, as Watson shows, they are inevitably entangled within conflictual relations of culture, history, and power. As the development project failed to consider this wider field within which the specific "cultural" institutions were embedded, so too were deep flaws introduced into the project's management, decision-making, and long-term sustainability.

Culture defined in relation to a specific "ethnic" group has also been put to developmentally oriented ends in the Andes where indigenous culture has been celebrated. As Sarah Radcliffe and Nina Laurie show in their chapter, indigenous development that treats diverse Indian ways of life as the launch pad for development, rather than as an obstacle, represents a significant reworking of the region's cultural economy and development. Self-generated indigenous development comes about when Indian groups can carve out a livelihood for themselves with secure access to resources, a setting for inter-cultural relations, and an enduring sense of cultural identity. However, indigenous development projects devised under the banner of neo-liberal agendas of free markets and multicultural good governance have not addressed this combination of requirements, emphasizing instead the ways in which Indians can self-produce themselves as cultural consumption items (*via* crafts, tourism, and services). A tension continues to exist therefore between the promise of indigenous development (demanded by politically astute and powerful indigenous confederations) and the experiences on the ground.

Another arena through which culture has been put to work in development has been in relation to the concept of social capital, and its supposed corollary, local development groups. As Gina Porter and Fergus Lyon discuss in their chapter here, the tendency for development policy to persist with its model of social capital = groups = development persists, regardless of the experience on the ground. Moreover, development projects implement subtle arrangements of credit availability, technical support, and appropriate transport technology yet, when it comes to culture, rely upon a nation-state model of "Ghanaian culture" that fails to engage sufficiently with the country's regional unevenness and its corresponding cultural, ecological and social diversity. Grassroots beneficiaries articulately express their understanding of their situation, and locate conflict at the heart of their understanding of the "culture" of groups.

These chapters examine in depth the ways in which classic forms of development intervention – targeted areas and beneficiaries in the majority world – bring in culture. The other chapters in this collection recast the frame for analysis of culture and development by exploring the nature of development as ongoing political economic dynamism in relation to culture. In other words, these chapters retain a focus on culture and its relationship to achievements of prosperity and other pursuits outside the parameters of development projects and programs.[1] As Jan Nederveen Pieterse shows in his chapter, long-run transformations in cultural economies have often contained within them interactions between different groups' cultures. In the context of trade exchanges over centuries, and in the context of culturally distinctive goods and services now available in the global North, cultural development is characterized by multicultural and multiracial interactions in globally reaching (and regionally meaningful) networks and flows. In such a context, Nederveen Pieterse argues, the notion of an ethnic economy is highly problematic as a conceptual and policy tool as it misses out the very fluidity and interactivity that these situations rest upon. His chapter reminds us of the importance of a historically (and – I would add – geographically) informed view of cultural relations.

One example of such global flows and multicultural interactions is explored in Michelle Bigenho's chapter on Bolivian musicians touring Japan. Bigenho describes how the nightly production of "Bolivian" music on stage in front of Japanese audiences rests upon the flexible and dynamic skills of inter-cultural interaction that Nederveen Pieterse describes for numerous other groups. Yet with post-Fordist political economies and neo-liberal imperatives of labor "flexibility" and entrepreneurialism, the conditions for wealth generation and secure livelihoods for the Bolivian musicians remain constrained. Moreover, the terms of power between representatives of an impoverished majority world nation and a wealthy country work to reproduce cultural stereotypes that perpetuate their uneven terms of engagement. Although the Bolivians attempt to recast Japanese audiences' knowledge of their country, their contracts and conditions of work limit their agency in this regard.

Al James in his chapter examines the experiences of Mormon high-tech firms in the United States, questioning the generalities about "culture" in much economic and economic geography literature. A detailed analysis of firms' behavior, work cultures, and priorities demonstrates that Mormon firms do not replicate exactly the "cultural" features expected in successful regional industrial clusters. Moreover, he shows how Mormon culture – conceptualized as a multi-scalar interaction between firms, industrial sectors, and regional cultures – establishes certain features of greater social sustainability (such as work–life balance and holidays). The case of these firms illustrates the diverse ways in which the pursuit of accumulation is simultaneously the pursuit of many other goals (Amin and Thrift 2004).

Finally, Cheryl McEwan's chapter provides an analysis of the memory cloths by South African women as an example of how culture engages issues of memory, power, and conflict. In the multicultural and post-conflict situation in South Africa, women turn to recording the horrors of the apartheid era with cloth and embroidery. Although these cultural productions cannot be romanticized as a therapeutic closure on apartheid in a celebratory postcolonial moment (Gooder and Jacobs 2002), McEwan highlights the women's direct engagement with legacies of violence, HIV status, and rape. Rather than being associated with processes of economic accumulation, such cultural work represents a strategic intervention in the accumulation of mutual respect, identity, and communication.

All of the chapters treat development in a globalized field of multiple meanings and practices where cultural difference – often the premise of previous development thinking and its current cultural turn – is embedded in a contested discursive and material arena. Culture is not simply an attribute of societies undergoing development but development itself operates as a cultural process. When development introduces "culture," it engages not only its own paradigm-specific working definition of culture (held variously, and sometimes at odds, by planners, practitioners, and beneficiaries), but also a broader field of different cultural registers (which include state and popular cultures, youth cultures, as well as overlapping structures of feelings). The chapters moreover highlight the need for a critical attitude to contested material culture and the broader processes of accumulation within which it is embedded. In the case studies of cultural development, the instrumental use of culture comes across clearly whether in relation to natural resource management, transport policy, or socioeconomic development. This critique of cultural development has been made before, yet can be pushed further in order to attempt a rethinking of the institutional and global context within which culture and development occurs.

Culture and development: culture as production, institution, and creativity in development[2]

In much recent culture and development thinking, culture has been brought into development primarily as a product and as an institution. Recent culture and development thinking draws explicitly on culture as a product and as an institution, and hence treats culture as a resource although the treatment of culture as product or institution entails certain consequences as detailed in the preceding chapters. As development appropriates culture as a resource or as an institution, each understanding entails a particular articulation of tradition and modernity, while each interpretation offers distinct templates for development interventions.

By treating culture as a product, culture and development projects can emphasize a product or service that is culturally distinctive, or which is associated with a simpler or healthier lifestyle. Development thinking has gone through two moves to reach this conclusion. First, there is a broad recognition that cultural activities and objects are economically remunerative. Second, income generation depends directly and indirectly upon cultural facilities and the general cultural environment (Sen 2004 for an overview; James, this volume). In relation to the first point, neo-liberal development projects explicitly aim to generate income from indigenous culture, for example (Davis 2002; Bigenho, this volume; Radcliffe and Laurie, this volume). Development and state officials alike recognize the attractive niche represented by non-Western lifestyles and products in which a romantic association with Indian lifeways contributes to a product/service's value and attractiveness (Kaplan 1995). Culture-as-product treats culture as a set of material objects and distinctive behaviors. When this interpretation of culture is inserted into development thinking, it promotes the orientation of culturally distinctive products and services onto the market. These interpretations are embedded within neo-liberal development's awareness of a post-Fordist global economy (Comoroff and Comoroff 2000), where in globalized markets for niche products in late modern political economies, culturally distinctive craft products appeal to the Western search for authentic lifestyles. In other words, cultural development thinking needs to be understood within the context of a globalizing capitalist market economy, rather than as – or in addition to experiments of – a grassroots "alternative" development model. Recognizing how global political economies work through difference, culture-as-product plays to the market opportunities opened up for culturally unique goods and services. Culture-as-product thus represents neither an alternative to globalization nor the protection of pristine unique cultures (and hence works against the critiques of globalization that inform much culture and development thinking).

In its emphasis on material culture, culture-as-product tends to suggest that the "traditional" or "classic" forms of culture are more appropriate for the market than hybrid or modern styles. Ethno-tourists wish to buy pottery or textiles, not plastic or synthetic fabrics. By ignoring cultural hybridity and dynamic material cultures, this policy prescription rests upon a problematic nostalgia for a clearly defined culture. Moreover, such development thinking reveals a strong anti-statist thread; the market is perceived as the arena where cultural diversity might be preserved. Whereas under modernization the invention of tradition by non-state actors was *in*authentic, in today's culture and development culture-as-product is seen to originate with *non-state* groups whose authenticity rests precisely on a perceived distance from the state.

Treating culture as an institution results in different development dynamics. In this aspect of development thinking, culture appears as a form of organization that provides structure and stability to society. During the 1990s, economic development planners and scholars expressed increasing concern about how non-economic factors (that is, factors unrelated to prices) influenced peoples' behavior. For example, development economics struggled to understand why sharecropping and other "pre-capitalist" economic behaviors were re-emerging in Indian agriculture (Hart 2002). New institutional economics, as this field became known, "can be seen as a way of trying to incorporate culture into orthodox economic theory" (Worsley 1999: 38). With the social stability associated with embedded cultural features, a number of projects draw upon so-called "traditional" forms of social organization in order to implement projects and to structure project management. Influenced by the concept of social capital, development policy seeks out the well-established and locally legitimate forms of social organization and decision-making to provide a facilitative arena within which development interventions can be initiated and implemented.

When development calls upon culture as an institution, culture combines the idea of stability and incorruptibility associated with a form of tradition that nevertheless connotes a modernist order. As a means of legitimating authority and local power relations, culture-as-institution appears to offer templates for the regulation of social groups in diverse fields such as natural resource management and credit organization. In effect, when mobilized for development interventions culture represents a non-state formality which carries considerable legitimacy in multilateral and agency thinking to promote civil society and participation under neo-liberal "good governance" policies to encourage *organized* civil society (Leftwich 1996). In the context of policies to establish neo-liberal governmentality such as the rolling-back of the state and the granting of legitimacy to non-state institutions (Shore and Wright 1997), culture holds out the promise of a non-state structure.

A different possible interpretation for cultural development is to treat culture as creativity, as a way of "thinking outside the box." Although this application of culture has been less frequently analyzed (compare Appadurai 2004), viewing culture as flexible innovation releases development from the disempowering forcing of cultures into markets or forms of political engineering. Culture and development thinking has only begun to examine and explore the potential and implications of viewing culture as creativity. When recognized at all, culture-as-creativity has been pictured as existing in non-state groups whose deployment of culture challenges and reconfigures political and economic relations. The cases of participatory budgeting in Porto Alegre, Brazil, and shantytown organizations in Villa El Salvador, Peru, have both been offered as examples of how grassroots social actors rework tradition with positive development outcomes (Kliksberg 1999). Similarly, the non-governmental Inter-American Foundation worked widely from the 1980s to release what it termed the "cultural energy" of grassroots popular culture throughout Latin America and the Caribbean (Kleymeyer 1995).

Projects on the ground offer opportunities to ensure agendas of intercultural understanding and empowerment by providing a flexible combination of tradition and modernity. In other words, culture can speak to adaptable political economies and open forms of governance, in which tradition or distinctiveness is not held to a rigid template but reflects the everyday *bricolage* and hybridity of livelihoods and forms of organization. Compared with anti-statist product and quasi-statist institution, culture-as-creativity arguably offers a less tradition-bound notion of cultural difference, in which engagements with facets of modernity are not a problem but part of the context of reworking "tradition." Defining culture as a flexible resource offers a way of drawing on social structures and meanings to offer innovative solutions – often in combination with existing social organization – to development problems. Nevertheless, culture-as-creativity requires further analysis in order to understand how culture acts as a toolbox for lateral thinking and empowering action. To highlight the potentially creative and problem-solving aspect of dynamic culture is not to sideline economic injustice or exclusionary racial formations that often compound uneven development outcomes in the global South. Treating culture as creativity entails dealing simultaneously with the structural inequalities of political economies, and the damaging effects of racial formations on subalterns' development opportunities.

*

Culture and development currently works from cultural *difference*, not from culture. The background cultural assumptions that informed much mainstream development thinking have been criticized roundly for their positionality, hierarchies, and exclusions. Such a critique has informed development's cultural turn. Yet by identifying cultural difference within the beneficiary population – a cultural difference that leaves unexamined the culture of the development agent – development leaves out the meaningful ground upon which cultural development is constructed and operationalized. Cultural difference is identified in the institutions, livelihood strategies, gender relations, and material cultures of beneficiary populations and is treated as another tool of development. In its appropriation as governance, product, or resource, culture *qua* cultural difference contributes to the managerialist and organizational goals of development, albeit with at times beneficial socioeconomic outcomes for participants. The reduction of culture to cultural difference reinscribes development's inability to deal with culture as creativity, by failing to take on board dynamism and fluid boundaries.

Yet the instrumentalization of culture exists itself within a contested cultural economy within which diverse actors are attempting to rework the bases of livelihood, politics and expression. In this context, culture and development refers to the fact that culture is not primordial but is reworked and reproduced around and through development, just as development (as political economy and as planned intervention) is embedded in "imaginaries of desirability," material culture and social relations. Treating culture and development as co-producing and recognizing the cultural imperatives to livelihood improvements offers a constructive way forward for development thinking and practice.

Notes

1   As Al James notes however (this volume), policy-makers in many countries are trying to devise projects that would replicate the success of Silicon Valley.

2   This section appears in a slightly amended form in Radcliffe and Laurie (2006).

Bibliography

Abu-Lughod, L. (1999) "Comments on C. Brumann," *Current Anthropology*, 40, supplement: S13–15.

Agarwal, A. (2001) "Common Property Institutions and Sustainable Governance of Resources," *World Development*, 29 (10): 1649–72.

Agosín, M. (1987) *Scraps of Life*, Trenton, NJ: Red Sea Press.

—— (1996) *Tapestries of Hope, Threads of Love: The Arpillera Movement in Chile, 1974–1994*, Alberquerque, NM: University of New Mexico Press.

Agrawal, A. (2005) *Environmentality*, Durham, NC: Duke University Press.

—— and Gibson, C. C. (1999) "Enchantment and Disenchantment: the Role of Community in Natural Resource Management," *World Development*, 27 (4): 629–50.

Aguilera, E. (2000) "Minority Owned Firms Lack in Backing," *Orange County Register*, September 25.

Albó, X. (1994) "And from Kataristas to MNRistas: the Surprising and Bold Alliance Between Aymaras and Neoliberals in Bolivia," in D. L. Van Cott (ed.) *Indigenous Peoples and Democracy in Latin America*, New York: St. Martin's Press, pp. 55–82.

Alexander, J. and Mohanty, C. T. (eds) (1997) *Feminist Genealogies, Colonial Legacies, Democratic Futures*, London: Routledge.

Alkire, S. (2004) "Culture, Poverty and External Intervention" in V. Rao and M. Walton (eds) *Culture and Public Action: A Cross-disciplinary Dialogue on Development Policy*, Stanford, CA: Stanford University Press, pp. 185–209.

Allen, J. (2003) *Lost Geographies of Power*, London, Blackwell.

Allen, T. (2000) "Taking Culture Seriously," in T. Allen and A. Thomas (eds) *Poverty and Development into the 21st Century*, 2nd edn, Oxford: Open University with Oxford University Press, pp. 443–68.

Amin, A. (1994) "Post-Fordism: Models, Fantasies and Phantoms of Transition," in A. Amin (ed.) *Post-Fordism: A Reader*, Oxford, and Cambridge, MA: Blackwell, pp. 1–40.

—— and Thrift, N. (1993) "Globalization, Institutional Thickness and Local Prospects," *Revue d'Économie Régionale et Urbaine*, 3: 405–27.

—— and Thrift, N. J. (2004) "Introduction," in A. Amin and N. J. Thrift (eds) *The Blackwell Cultural Economy Reader*, Oxford: Blackwell, pp. x–xxx.

Amponsem, G. (1996) "Global Trading and Business Networks Among Ghanaians: an Interface of the Local and the Global," Ph.D. dissertation, Bielefeld University.

Anchirinah, V. M. and Yoder, R. (2000) "Evaluation of the Pilot Phase of the Intermediate Means of Transport (IMT) of the Village Infrastructure Project in Some Selected Districts of the Brong-Ahafo and Ashanti Regions." Prepared for SelfHelp Foundation, Kumasi, September 2000.

Anderson, B. (1983) *Imagined Communities*, London: Verso.

Anderson, K., Domosh, M., Pile, S., and Thrift, N. (eds) (2002) *Handbook of Cultural Geography*, London: Sage.

Anderson, P. (2000) "Renewals," *New Left Review*, (second series) 1: 5–24.

Andolina, R., Laurie, N., and Radcliffe, S. A. (forthcoming) *Multiethnic Transnationalism: Indigenous Development in the Andes*, London: Duke University Press.

——, Radcliffe, S. A., and Laurie, N. (2005) "Development and Culture: Transnational Identity Making in Bolivia," *Political Geography*, 24 (6): 678–702.

Andrejczak, M. (1999) "In L.A. Korean Area, 7 Banks Are Too Many," *The American Banker* (September 20).

Antara News Agency (1999) "Indonesian Immigrants in US Lag Behind Others" (September 21).

Annan, K. (1998) "The Causes of Conflict and the Promotion of Durable Peace and Sustainable Development in Africa," report of the UN Secretary-General. (http://www.theire.org/index.cfm/www10/2132 (accessed May 2005.)

Anon. (2001) "Healing the Heart with Art," *Daily News* (Durban), Tuesday, May 29, 7.

Apffel-Marglin, F. (1998) *The Spirit of Regeneration: Andean Culture Confronting Western Notions of Development*, London: Zed Books.

Appadurai, A. (1996) "Disjuncture and Difference in the Global Cultural Economy," in J. X. Inda and R. Rosaldo (eds) *The Anthropology of Globalization: A Reader*, Oxford: Blackwell, pp. 46–64.

—— (2004) "The Capacity to Aspire: Culture and the Terms of Recognition" in V. Rao and M. Walton (eds) *Culture and Public Action: A Cross-disciplinary Dialogue on Development Policy*, Stanford, CA: Stanford University Press, pp. 59–84.

Apter, D. (2005) *The Pan African Nation*, Chicago: University of Chicago Press.

Arce, A. and Long, N. (eds) (2000) *Anthropology, Development and*

Modernities: Exploring Discourses, Counter-Tendencies and Violence, London: Routledge.

Archer, B. (1997) "A Patchwork of Internment," *History Today*, July: 11–18.

Arizpe, L. (2004) "The Intellectual History of Culture and Development Institutions," in V. Rao and M. Walton (eds) *Culture and Public Action*, Stanford, CA: Stanford University Press, pp. 163–84.

Arrington, L. J. and Bitton, D. (1992) *The Mormon Experience: A History of the Latter-Day Saints* (2nd edn), Chicago: University of Illinois Press.

Asheim, B. T. (1996) "Industrial Districts as 'Learning Regions': a Condition for Prosperity?," *European Planning Studies*, 4: 379–400.

—— (2001) "The Learning Firm in the Learning Region: Broad Participation As Social Capital Building," paper presented at the *Annual Conference of the Association of American Geographers*, New York, February 27– March 3.

—— and Cooke, P. (1999) "Local Learning and Interactive Innovation Networks in a Global Economy," in E. J. Malecki and P. Oinas (eds) *Making Connections: Technological Learning and Regional Economic Change*, Aldershot: Ashgate Publishing, pp. 145–78.

Asmal, K. (2000) "Truth, Reconciliation and Justice the South African Experience in Perspective," *The Modern Law Review*, 63 (1): 9–10.

Assies, W., van der Haar, G., and Hoekema, A. (eds) (2001) *The Challenge of Diversity: Indigenous Peoples and Reform of the State in Latin America*, Amsterdam: Thela thesis.

Atak, I. (1999) "Four Criteria of Development NGO Legitimacy," *World Development*, 27 (5): 855–64.

Atingdui, L., Anheier, H. K., Sokolowski, S. W., and Laryea, E. (1998) "The Non-Profit Sector in Ghana," in H. K. Anheier and L. M. Salamon (eds) *The Nonprofit Sector in the Developing World*, Manchester: Manchester University Press, pp. 158–97.

Atkins, A. and Rey-Maquieira, E. (1996) *Ethno-Development: a Proposal to Save Colombia's Pacific Coast*, London: CIIR.

Atkins, E. (2001) *Blue Nippon: Authenticating Jazz in Japan*, Durham, NC and London: Duke University Press.

Bahr, S. J. (1994) "Religion and Adolescent Drug Use: a Comparison of Mormons and Other Religions," in M. Cornwall, T. B. Heaton, and L. A. Young (eds) *Contemporary Mormonism: Social Science Perspectives*, Chicago: University of Illinois Press, pp. 118–37.

Bank Information Center (2004) *Indigenous Peoples and the World Bank*, Bank Information Center website, accessed October 27, 2004.

Barnard, A. and Kenrick, J. (eds) (2001) *Africa's Indigenous Peoples: "First Peoples" or "Marginalized Minorities"?*, Edinburgh: Centre of African Studies, University of Edinburgh.

Barnes, T. (2002) "Introduction: 'Never Mind the Economy. Here's Culture,'" in K. Anderson, M. Domosh, S. Pile, and N. Thrift (eds) *Handbook of Cultural Geography*, London: Sage, pp. 89–97.

Barnett, C. (1998) "Cultural Twists and Turns", *Environment and Planning D*, 16: 631–4.

Barr, A. and Toye, J. (2000) "It's Not What You Know – It's Who You Know! Economic Analysis of Social Capital" ID21 *Development Research Insights* September (2000), 34.

Barry, A., Osborne, T., and Rose, N. (1993) "Liberalism, Neoliberalism, and Governmentality: an Introduction" *Economy and Society*, 22: 265–6.

Bassi, M. (1997) "Returnees in Moyale District, Southern Ethiopia: New Means for an Old Inter-ethnic Game," in R. Hogg (ed.) *Pastoralists, Ethnicity and the State in Ethiopia*, London: Haan, pp. 23–54.

Baxter, P. T. W., Hultin, J., and Triulzi, A. (1996) "Introduction," in P. T. W. Baxter, J. Hultin, and A. Triulzi (eds) *Being and Becoming Oromo: Historical and Anthropological Enquiries*, Lawrenceville, NJ: Red Sea Press, pp. 7–25.

Bebbington, A. (2002) "Sharp Knives and Blunt Instruments: Social Capital in Development Studies," *Antipode* 34 (4): 800–3.

——, Guggenheim, S., Olson, E., and Woolcock, M. (2004) "Exploring Social Capital Debates at the World Bank," *Journal of Development Studies*, 40 (5): 33–64.

Beck, U. (2000) *What Is Globalization?*, Oxford: Blackwell.

Bennett, J. (2000) "The Politics of Writing," *Agenda*, 46: 3–12.

Berger, M. T. (2003) "The New Asian Renaissance and Its Discontents: National Narrative, Pan-Asian Visions and the Changing Post-Cold War Order," *International Politics*, 40 (2): 195–222.

Berlant, L. (1991) *The Anatomy of National Fantasy*, Chicago: University of Chicago Press.

Berman, M. (1980) *All That Is Solid Melts into Air*, New York, Harper.

Besteman, C. (1996) "Representing Violence and 'Othering' Somalia," *Cultural Anthropology*, 11 (1): 120–33.

Bhabha, H. (1992) "Postcolonial Criticism" in S. Greenblatt and G. Gunn (eds) *Redrawing the Boundaries: English and American Studies*, New York: MLA, 437–65.

Bhavnani, K.-K., Foran, J., and Kurian, P. (2003) "An Introduction to Women, Cutlure and Development," in K.-K. Bhavnani, J. Foran, and P. Kurian (eds) *Feminist Futures. Re-imagining Women, Culture and Development*, London: Zed Books, pp. 1–21.

Bigenho, M. (2002) *Sounding Indigenous: Authenticity in Bolivian Music Performance*, New York: Palgrave-Macmillan.

—— (in press) "Making Music Safe for the Nation: Folklore Pioneers in Bolivian Indigenism," In *Natives Making Nation: Gender, Identity and Indigeneity in the Andes*, A. Canessa (ed.) Tucson, AZ: Arizona University Press.

Black, P. A. (2002) "On the Case for 'Black Economic Empowerment' in South Africa," *South African Journal of Economics*, 70 (8): 1148–62.

Black, R. and Watson, E. (2006) "Local Community, Legitimacy and Cultural Authenticity in Post-Conflict Natural Resource Management," *Environment and Planning D: Society and Space.*

Blunt, A. and McEwan, C. (eds) (2002) *Postcolonial Geographies*, London: Continuum.

Boku Tache and Irwin, B. (2003) "Traditional Institutions, Multiple Stakeholders and Modern Perspectives in Common Property: Accompanying Change Within Borana Pastoralist Systems," *Securing the Commons*, no. 4, pp. 1–51.

Bonacich, E. and Applebaum, R. P. (2000) *Behind the Label: Inequality in the Los Angeles Apparel Industry*, Berkeley, CA: University of California Press.

Bondi, L. and Laurie, N. (eds) (2005) "Special Issue: Working the Spaces of Neo-Liberalism: Activism, Professionalization and Incorporation" *Antipode*, 37 (3).

Bonnell, V. and Hunt, L. (eds) (1999) *Beyond the Cultural Turn*, Berkeley, CA: University of California Press.

Boubakri, H. (1999) "Les Entrepreneurs Migrants d'Europe: Dispositifs Communautaires et Économie Ethnique; le Cas des Entrepreneurs Tunisiens en France," *Cultures et Conflits*, 33/34: 69–88.

Bourdieu, P. (1976) "Les Modes de domination," *Actes de la Recherche en Sciences Sociales*, 2 (2/3): 122–32.

—— (1977) *A Theory of Practice*, Cambridge: Cambridge University Press.

—— (1980) "Le Capital Sociale," *Actes de la Recherche en Sciences Sociales*, 31: 2–3.

Bozzoli, B. (1998) "Public Ritual and Private Transition: the Truth Commission in Alexandra Township, South Africa 1996," *African Studies*, 57 (2): 167–95.

Brenner, R. (2002) *The Boom and the Bubble: the US in the World Economy*, London: Verso.

Brink, A. (1998) "Stories of History: Reimagining the Past in Post-apartheid Narrative" in S. Nuttall and C. Coetzee (eds) *Negotiating the Past: The Making of Memory in South Africa*, Oxford: Oxford University Press, pp. 29–42.

Brohman, J. (1996) *Popular Development*, Oxford: Blackwell.

Brown, D. L. and Ashman, D. (1999) "Social Capital, Mutual Influence, and Social Learning in Intersectoral Problem Solving in Africa and Asia," in D. L. Cooper-Rider and J. E. Dutton (eds) *Organizational Dimensions of Global Change: No Limits to Cooperation*, London: Sage, pp. 139–67.

Brysk, A. (2000) *From Tribal Village to Global Village: Indian Rights and International Relations in Latin America*, Stanford, CA: Stanford University Press.

Buck-Morse, S. (2003) *Thinking Past Terror*, London: Verso.

Bulbeck, C. (1998) *Re-Orienting Western Feminism: Women's Diversity in a Postcolonial World*, Cambridge: Cambridge University Press.

Burawoy, M. (2003) "For a Sociological Marxism: the Complementary Converge of Antonio Gramsci and Karl Polanyi," *Politics and Society*, 31 (2): 193–261.

——, Blum, J., and George, S. *et al.* (eds) (2000) *Global Ethnography: Forces, Connections and Imaginations in a Postmodern World*, Berkeley, CA: University of California Press.

Calla, R. (2001) "Indigenous Peoples, the Law of Popular Participation and Changes in Government, Bolivia, 1994–98" in W. Assies, G. van der Haar, and A. Hoekema (eds) *The Challenge of Diversity*, Amsterdam: Thela thesis, pp. 77–94.

Camagni, R. (ed.) (1991) *Innovation Networks: Spatial Perspectives*, London: Belhaven Press.

Capello, R. (1999) "Spatial Transfers of Knowledge in High Technology Milieux: Learning versus Collective Learning Processes," *Regional Studies*, 33 (4): 353–65.

Carranza, J. (2002) "Culture and Development: Some Considerations for Debate," *Latin American Perspectives*, issue 125, 29 (4): 31–46.

Castells, M. and Hall, P. (1994) *Technopoles of the World: The Making of the Twenty-First Century Industrial Complexes*, London: Routledge.

Chambers, R. (1997) *Whose Reality Counts? Putting the First Last*, London: IT Press.

Chang, J. H. Y. (2003) "Culture, State and Economic Development in Singapore," *Journal of Contemporary Asia*, 33 (1): 85–105.

Chari, S. (2004) *Fraternal Capital*, Stanford, CA: Stanford University Press.

Chen, Y. (2000) *Chinese San Francisco, 1850–1943: A Trans-Pacific Community* G. Chang (ed.), Stanford, CA: Stanford University Press.

Ching, L. (2001) "Globalizing the Regional, Regionalizing the Global: Mass Culture and Asianism in the Age of Late Capital," in A. Appadurai (ed.) *Globalization*, Durham, NC and London: Duke University Press, pp. 279–306.

Chirwa, W. (1997) "Collective Memory and the Process of Reconciliation and Reconstruction," *Development in Practice*, 7 (4): 479–82.

Choenni, A. (2000) *Bazaar in de Metropool: Allochtone Detailhandel in Amsterdam en Achtergronden van Haar Locale Begrenzing*, Amsterdam: Emporium.

Chua, P., Bhavnani, K., and Foran, J. (2000) "Women, Culture and Development: A New Paradigm for Development Studies," *Ethnic and Racial Studies*, 23 (5): 820–41.

Church of Jesus Christ of Latter-day Saints/Deseret News, The (2000) *2001–2002 Church Almanac*, Salt Lake City, UT: Deseret News.

Cia, B. and Martí, J. (2004) "El Reto de la Agenda 21 de la Cultura: Contra el malestar," *El País* (Madrid), May 8, 2004, pp. 2–3.

Clague, C. and Grossbard-Shechtman, S. (eds) (2001) "Special Issue: Culture and Development: International Perspectives," *Annals of the American Academy of Political and Social Science*, vol. 573, January.

Cleaver, F. (2001) "Institutions, Agency and the Limitations of Participatory Approaches to Development," in B. Cooke and U. Kothari (eds) *Participation: the New Tyranny?*, London: Zed Books, pp. 36–55.

Clifford, M. (2001) *Political Genealogy after Foucault*, London: Routledge.

Cobas, J.A. (1989) "Six Problems in the Sociology of the Ethnic Economy," *Sociological Perspectives*, 32 (2): 201–14.

Cohen, D. and Prusak, L. (2001) *In Good Company. How Social Capital Makes Organizations Work*, Boston, MA: Harvard Business School Press.

Cohen, W. M. and Levinthal, D. A. (1990) "Absorptive Capacity: a New Perspective on Learning and Innovation," *Administrative Science Quarterly*, 35 (1): 128–52.

Coleman, J. S. (1988) "Social Capital in the Creation of Human Capital," *American Journal of Sociology*, 94: 95–120.

Comaroff, J. and Comaroff, J. L. (2000) "Millennial Capitalism: First Thoughts on a Second Coming," *Public Culture*, 12 (2): 291–343.

Condry, I. (2001) "Japanese Hip Hop and the Globalization of Popular Culture," in G. Gmelch and W. Zenner (eds) *Urban Life*, 4th edn, Prospect Heights, IL: Waveland Press, pp. 357–87.

Cone, C. (1995) "Crafting Selves: the Lives of Mayan Women," *Annals of Tourism Research*, 22 (2): 314–27.

Cooke, B. and Kothari, U. (eds) (2001) *Participation: A New Tyranny?*, London: Zed Books.

Cooper, F. (2005) *Colonialism in Question*, Berkeley, CA: University of California Press.

—— and Packard, R. (eds) (1997) *International Development and the Social Sciences*, Berkeley, CA: University of California Press.

Corbridge, S. (1998) "Review of Escobar 1995," *Journal of Development Studies*, 34 (6): 138–48.

Cornwall, M. (2001) "Toward a Sociological Analysis of Mormonism," in M. Cornwall, T. B. Heaton, and L. A. Young (eds) *Contemporary Mormonism: Social Science Perspectives*, Urbana, IL and Chicago: University of Illinois Press, pp. 1–9.

Coronil, F. (1999) *The Magical State*, Chicago: University of Chicago Press.

Cosgrove, D. E., and Jackson, P. (1987) "New Directions in Cultural Geography," *Area*, 19 (2): 95–101.

Cowen, M. and Shenton, R. (1997) *Doctrines of Development*, London: Routledge.

Crain, M. (1996) "The Gendering of Ethnicity in the Ecuadorian Andes: Native Women's Self-Fashioning in the Urban Marketplace," in M. Melhuus and K. A. Stolen (eds) *Machos, Mistresses, Madonnas: Contesting the Power of Latin American Gender Imagery*, London: Verso, pp. 134–58.

Crang, P. (1997) "Introduction: Cultural Turns and the (Re)Constitution of Economic Geography," in R. Lee and J. Wills (eds) *Geographies of Economies*, London: Arnold, pp. 3–15.

Crush, J. (ed.) (1995) *The Power of Development*, London: Routledge.

Curtin, P. D. (1984) *Crosscultural Trade in World History*, Cambridge: Cambridge University Press.

d'Andrea, L., d'Arca, R., and Mezzana, D. (1998) *Handbook on the Social and Economic Integration Practices of Immigrants In Europe*, Rome: CERFE.

Davis, M. (2005) *Planet of the Slums*, London, Verso.

Davis, S. (1999) "Bringing Culture into the Development Paradigm: the View from the World Bank," *Development Anthropologist*, 16 (1–2): 25–31.

—— (2002) "Indigenous Peoples, Poverty and Participatory Development: the Experience of the World Bank in Latin America," in R. Sieder (ed.) *Multiculturalism in Latin America: Indigenous Rights, Diversity and Democracy*, London: Palgrave-Macmillan, pp. 227–51.

Deal, T. E. and Kennedy, A. A. (2000) *Corporate Cultures: The Rites and Rituals of Corporate Life*, Cambridge, MA: Perseus Publishing.

Dean, M. (1991) *The Constitution of Poverty*, London: Routledge.

—— (1999) *Governmentality*, London: Sage.

DeFilippis, J. (2002) "Symposium of Social Capital: an Introduction," *Antipode*, 34 (4): 790–5.

de Kok, I. (1998) "Cracked Heirlooms: Memory on Exhibition," in S. Nuttall and C. Coetzee (eds) *Negotiating the Past: The Making of Memory in South Africa*, Oxford: Oxford University Press, pp. 57–74.

dei Ottati, G. (1994) "Trust, Interlinking Transactions and Credit in the Industrial District," *Cambridge Journal of Economics*, 18 (6): 529–46.

Dennis, C. and Peprah, E. (1995) "Coping with Transition Through Organisation: Techiman Market, Ghana," *Gender and Development*, 3 (3): 43–8.

Department for International Development (2002) *Ghana: Country Strategy Paper*, London: Department for International Development, November.

Department of Trade Industry, HM Government (1998) *Biotechnology Clusters*, report of a team led by Lord Sainsbury, Minister for Science, London: HMSO, August.

Deruyttere, A. (1997) *Indigenous Peoples and Sustainable Development: The Role of the Inter-American Development Bank*, Washington, DC: Inter-American Development Bank.

De Soto, H. (1989) *The Other Path: the Invisible Revolution in the Third World*, New York: Harper & Row.

Deutsche Presse-Agentur (2000) "Mysterious Chinese Immigrants Baffle the Spaniards," *Hindustan Times* (July 30).

De Weerdt, J. (2001) "Community Organisations in Rural Tanzania: a Case Study of the Community of Nyakatoke Bukoba Rural District," IDS, University of Dar es Salaam, Tanzania, and CES, University of Leuven, Belgium.

Dezalay, Y. and Garth, B. (2002) *The Internationalization of Palace Wars: Lawyers, Economists and the Contest to Transform Latin American States*, Chicago: University of Chicago Press.

Díaz, P. (2005) "Licencia para Curar: el Parlamento de Súrafrica Aprueba Normas para Regular la Medicina de los Antepasados," *El País* [Madrid], April 23, p. 72.

DiBella, A. J., Nevis, E. C., and Gould, J. M. (1996) "Organisational Learning Style As a Core Capability," in B. Moingeon and A. Edmondson (eds), *Organisational Learning and Competitive Advantage*, London: Sage, pp. 38–55.

di Leonardo, M. (1984) *The Varieties of Ethnic Experience: Kinship, Class, and Gender among California Italian-Americans*, Ithaca, NY: Cornell University Press.

Dobbin, C. (1996) *Asian Entrepreneurial Minorities: Conjoint Communities in the Making of the World-economy, 1570–1940*, Richmond, Surrey: Curzon Press.

Donham, D. L. (1999) *Marxist Modern: an Ethnographic History of the Ethiopian Revolution*, Berkeley, CA: University of California Press; Oxford: James Currey.

—— and James, W. (eds) (1986) *The Southern Marches of Imperial Ethiopia: Essays in History and Social Anthropology*, Cambridge: Cambridge University Press.

Drayton, R. (1996) *Nature's Government*, New Haven, CT: Yale University Press.

Duggan, L. (2003) *The Twilight of Equality*, Boston, MA: Beacon.

Dunn, E. (1996) "Money, Morality and Modes of Civil Society Among American Mormons", in C. Hann and E. Dunn (eds) *Civil Society: Challenging Western Models*, London: Routledge.

Duvenage, P. (1999) "The Politics of Memory and Forgetting After Auschwitz and Apartheid," *Philosophy and Social Criticism*, 25 (3): 1–17.

Eagleton, T. (2000) *The Idea of Culture*, Oxford: Blackwell.

Ehlers, T. B. (1990) *Silent Looms: Women and Production in a Guatemalan Town*, Boulder, CO: Westview Press.

Eisenberg, A. (1999) "Trust, Exploitation and Multiculturalism," unpublished paper, University of Victoria, BC.

Ellmeier, A. (2003) "Cultural Entrepreneurialism: on the Changing Relationship between the Arts, Culture and Employment," *International Journal of Cultural Policy*, 9 (1): 3–16.

Elyachar, J. (2002) "Empowerment Money: the World Bank, Non-governmental Organizations, and the Value of Culture in Egypt," *Public Culture*, 14 (3): 493–513.

Escobar, A. (1995) *Encountering Development. The Making and Unmaking of the Third World*, Princeton, NJ: Princeton University Press

—— (2001) "Culture Sits in Places: Reflections on Globalism and Subaltern Strategies of Localisation," *Political Geography*, 20: 139–74.

Evans, P. (1996) "Development Strategies Across the Public-Private Divide," *World Development*, 24 (6): 1033–7.

Eyben, R. (2000) "Development and Anthropology: a View from Inside the Agency," *Critique of Anthropology*, 20 (1): 7–14.

Fairhead, J. and Leach, M. (1996) *Misreading the African Landscape*, Cambridge: Cambridge University Press.

Favell, A. (2001) "Globalisation, Immigrants and Euro-elites: Questioning the Transnational Social Power of Migrants," unpublished paper.

Fedderke, J., de Kadt, R., and Luiz, J. (1999) "Economic Growth and Social Capital: A Critical Reflection," *Theory and Society*, 28: 709–45.

Feld, S. (2001) "A Sweet Lullaby for World Music," in A. Appadurai (ed.) *Globalization*. Durham, NC and London: Duke University Press, pp. 189–216.

Ferguson, J. (1994) *The Anti-Politics Machine: "Development," Depoliticization and Bureaucratic Power in Lesotho*, Minneapolis and London: University of Minnesota Press.

Field, S. (2001) "Remembering Experience, Interpreting Memory: Life Stories from Windermere," *African Studies*, 60 (1): 119–33.

Fine, B. (2001a) "Social Capital and the Realm of the Intellect," *Economic and Political Weekly* (March 3).

—— (2001b) *Social Capital versus Social Theory*, London: Routledge.

—— (2002a) "They F**k You Up Those Social Capitalists," *Antipode*, 34 (4): 796–9.

Finn, P. (2000) "The Immigration Imbroglio," *International Herald Tribune* (November 24).

Fisman, R. (2000) "Preferential Credit? Ethnic and Indigenous Firms Vie for Equal Access," *Development Research Insights*, 34: 2–3.

Florida, R. and Kenney, M. (1990) "Why Silicon Valley and Route 128 Won't Save Us," *California Management Review*, 33 (1): 68–88.

Forsyth, T. (2005) *Encyclopaedia of International Development*, London: Routledge.

Foucault, M. (1977) *Discipline and Punish*, New York: Harper.

—— (1991) "Govermentality," in G. Burchell, C. Gordon, and P. Miller (eds) *The Foucault Effect*, Chicago: University of Chicago Press, pp. 87–104.

—— (2000) *Power*, New York: Pantheon.

Fountain, J. E. (1997) "Social Capital: A Key Enabler of Innovation," in L. M. Branscomb and J. H. Keller (eds) *Investing in Innovation: Creating a Research and Innovation Policy That Works*, Boston, MA: MIT Press, pp. 85–111.

Fox, R. and King, B. (eds) (2002) *Anthropology beyond Culture*, London: Routledge.

Fraser, N. (1997) *Justice Interruptus: Critical Reflections on the Post-Socialist Condition*, New York: Routledge.

—— and Honneth, A. (2003) *Redistribution or Recognition?*, London: Verso.

Freidburg, S. E. (2001) "Gardening on the Edge: The Social Conditions of Unsustainability on an African Urban Periphery," *Annals of the Association of American Geographers*, 91 (2): 349–69.

Friedmann, J. (1992) *Empowerment: The Politics of Alternative Development*, Oxford: Blackwell.

Fukuyama, F. (1995) "Social Capital and the Global Economy," *Foreign Affairs*, 74 (5): 89–103.

Furnivall, J. S. (1939) *Netherlands India: A Study of Plural Economy*, Cambridge: Cambridge University Press.

Gambetta, D. (1988) "Can We Trust Trust?" in D. Gambetta (ed.) *Trust: Making and Breaking Cooperative Relations*, Oxford: Blackwell, pp. 213–37.

Gaonkar, D. (ed.) (2001) *Alternative Modernities*, Durham, NC: Duke University Press.

García Canclini, N. (1992) "Cultural Reconversion," in G. Yúdice, J. Franco, and J. Flores (eds) *On Edge: The Crisis of Contemporary Latin American Culture*, Minneapolis: University of Minnesota Press, pp. 29–43.

Gardner, K. and Lewis, D. (1996) *Anthropology, Development and the Post-Modern Challenge*, London: Pluto.

Gertler, M. S. (1995) "'Being There': Proximity, Organization and Culture in the Development and Adoption of Advanced Manufacturing Technologies," *Economic Geography*, 71: 1–26.

—— (1997) "The Invention of Regional Culture," in R. Lee and J. Wills (eds), *Geographies of Economies*, London: Arnold, pp. 47–58.

—— (2004) *Manufacturing Culture: The Institutional Geography of Industrial Practice*, Oxford: Oxford University Press.

Geschiere, P. and Nyamnjoh, F. (2000) "Capitalism and Autochthony: the Seesaw of Mobility and Belonging,*" Public Culture*, 12 (2): 423–52.

Getachew, K. (1996) "The Displacement and Return of Pastoralists in Southern Ethiopia: A Cast Study of the Garri," in T. Allen (ed.) *In Search of Cool Ground: War, Flight and Homecoming in Northeast Africa*, London: UNRISD; Oxford: James Currey, pp. 111–23.

Gianturco, P. and Tuttle, T. (2000) *In Her Hands: Craftswomen Changing the World*, London: Penguin.

Gibson-Graham, J. K. (1996), *The End of Capitalism (As We Knew It): A Feminist Critique of Political Economy*, Blackwell: Oxford.

Gill, L. (2000) *Teetering on the Rim: Global Restructuring, Daily Life, and the Armed Retreat of the Bolivian State*, New York: Columbia University Press.

Giroux, H. (2004) *The Terror of Neoliberalism*, London: Paradigm.

Glacken, G. (1967) *Traces on the Rhodian Shore*, Berkeley: University of California Press.

Gold, S. J. and Light, I. (2000) "Ethnic Economies and Social Policy," *Research in Social Movements, Conflicts and Change*, 22: 165–91.

Goldman, M. (2005) *Imperial Nature*, New Haven, CT: Yale University Press.

González, E. (2003) *La discriminación en Chile: el caso de las mujeres Mapuche*, Working paper 23. Ñuke Mapuforlaget, Temuco.

Gooder, H. and Jacobs, J. (2002) "Belonging and Non-Belonging: The Apology in a Reconciling Nation," in A. Blunt and C. McEwan (eds) *Postcolonial Geographies*, London: Continuum, pp. 200–13.

Goodman, R. (1999) "Culture as Ideology: Explanations for the Development of the Japanese Economic Miracle," in T. Skelton and T. Allen (eds) *Culture and Global Change*. London: Routledge, pp. 127–36.

Goswami, M. (2004) *Producing India*, Chicago: University of Chicago Press.

Gowan, P. (1999) *The Global Gamble*, London: Verso.

Graham, R. (ed.) (1990) *The Idea of Race in Latin America, 1870–1940*, Austin, TX: University of Texas Press.

Granovetter, M. (1985) "Economic Action and Social Structure: the Problem of Embeddedness," *American Journal of Sociology*, 91: 481–510.

—— (1994) "Business Groups," in N. J. Smelser and R. Swedberg (eds) *The Handbook of Economic Sociology*, Princeton, NJ: Princeton University Press, pp. 233–6.

Gray, M. and James, A. (2006) "Connecting Gender and Economic Competitiveness: Lessons from Cambridge's High Tech Economy," *Environment and Planning A.*

Graybill, L. (2001) "The Contribution of the Truth and Reconciliation Commission Toward the Promotion of Women's Rights in South Africa," *Women's Studies International Forum*, 24 (1): 1–19.

Grewal, I. and Kaplan, C. (eds) (1994) *Scattered Hegemonies: Postmodernity and Transnational Feminist Practices*, London: University of Minnesota Press.

Grief, A. (1994) "Cultural Beliefs and the Organization of Society: a Historical and Theoretical Reflection on Collectivist and Individualist Societies," *Journal of Political Economy*, 102 (5): 912–15.

Griffin, K. (1996) "Culture, Human Development and Economic Growth." Working paper in economics 96–17. Riverside, CA: University of California-Riverside.

—— (2000) "Culture and Economic Growth: The State and Globalization," in J. Nederveen Pieterse (ed.) *Global Futures*, London: Zed Books, pp. 189–202.

Griffiths, T. (2000) *World Bank Projects and Indigenous Peoples in Ecuador and Bolivia*, Paper for presentation in Washington, DC, May 2000. Washington DC: Forest Peoples Programme/Bank Information Center.

Grootaert, C. and Narayan, D. (2001) "Local Institutions, Poverty and Household Welfare in Bolivia." Washington, DC: World Bank.

Gufu, O. (1996) "Shifting Identities along Resource Borders," In P. T. W. Baxter, J. Hultin, and A. Triulzi (eds) *Being and Becoming Oromo: Historical and Anthropological Enquiries*, Lawrenceville, NJ: Red Sea Press, pp. 117–31.

Guijt, I. and Shah, M. (eds) (1998) *The Myth of Community: Gender Issues in Participatory Development*, London: ITDG Publishing.

Gupta, A. (1998) *Postcolonial Developments: Agriculture in the Making of Modern India*, Durham, NC and London: Duke University Press.

—— and Ferguson, J. (eds) (1997) *Culture, Power, Place: Explorations in Critical Anthropology*, Durham, NC: Duke University Press.

Gwynne, B. and Kay, C. (eds) (2004) *Latin America Transformed: Globalization and Modernity*, 2nd edn, London: Arnold.

Haberfellner, R. (2000) "Ethnische Ökonomien als Forschungsgegenstand der Sozialwissenschaften," *SWS Rundschau*, 40 (1): 43–62.

—— and Böse, M. (1999) "Ethnische Ökonomien," in H. Fassman (ed.) *Abgrenzen, Ausgrenzen, Aufnehmen. Empirische Befunde zu Fremdenfeindlichkeit und Integration*, Klagenfurt: Drava Verlag, pp. 75–94.

Hale, C. (2002) "Does Multiculturalism Menace? Governance, Cultural Rights and the Politics of Identity in Guatemala," *Journal of Latin American Studies*, 34: 485–524.

Hall, S., Hobson, D., Lowe, A., and Willis, P. (eds) (1980) *Culture, Media and Language: Working Papers in Cultural Studies 1972–1979*, London: Hutchinson.

Halter, M. (2000) "Chasing the Rainbow: Now that Marketers Realize People Come in Other Shades Besides White, Ethnic Background Is a Sizzling Commodity," *San Francisco Chronicle*, December 10.

Hannerz, U. (1992) *Culture, Cities and the World*, Amsterdam: Centre for Metropolitan Research.

—— (2002 [1989]) "Notes on the Global Ecumene," in J. X. Inda and R. Rosaldo, (eds) *The Anthropology of Globalization: A Reader*, Malden, MA and Oxford: Blackwell, pp. 37–45.

Hansen, T. and Stepputat, F. (eds) (2001) *States of Imagination: Ethnographic Explorations of the Postcolonial State*, Durham, NC: Duke University Press.

Harrison, L. and Huntington, S. (eds) (2000) *Culture Matters: How Values Shape Human Progress*, New York and London: Basic Books.

Harrison, M. I. (1994) *Diagnosing Organizations: Methods, Models and Processes*, Thousand Oaks, CA: Sage.

Harriss, J. (2002) *Depoliticizing Development: the World Bank and Social Capital*, London: Anthem Press.

Hart, G. (2001) "Development Critiques in the 1990s: Culs de Sac and Promising Paths," *Progress in Human Geography*, 25 (4): 649–58.

—— (2002) "Geography and Development: Development/s Beyond Neoliberalism? Power, Culture, Political Economy," *Progress in Human Geography*, 26: 812–22.

—— (2003) *Disabling Globalization*, Berkeley, CA: University of California Press.

Harvey, D. (1987) *The Urban Experience*, Baltimore: Johns Hopkins University Press.

—— (1989) *The Condition of Postmodernity: An Enquiry into the Origins of Cultural Change*, Oxford, UK and Cambridge, MA: Basil Blackwell.

—— (2002) *The New Imperialism*, London: Clarendon.

—— (2005) *A Brief History of Neoliberalism*, Oxford: Oxford University Press.

Healy, K. (1996) "Ethnodevelopment of Indigenous Bolivian Communities," in A. Kolata (ed.) *Tiwanaku and its Hinterland*, Washington DC: Smithsonian Institution Press, pp. 241–63.

—— (2000) *Llamas, Weavings and Organic Chocolate: Multicultural Grassroots Development Experience from the Andes and Amazon of Bolivia*, South Bend, IN: University of Notre Dame Press.

Heaton, T. B. (2001) "Mormon Families over the Life Course," paper presented to the University of Utah Population Seminar Group, January 19, 2001.

Heese, K. (2003) "Black Economic Empowerment in South Africa: a Case Study of Non-Inclusive Stakeholder Engagement," *Journal of Corporate Citizenship*, 12: 93–101.

Helland, J. (1996) "The Political Viability of Boorana Pastoralism." In P. T. W. Baxter, J. Hultin, and A. Triulzi (eds) *Being and Becoming Oromo: Historical and Anthropological Enquiries*, Lawrenceville, NJ: Red Sea Press, pp. 132–49.

—— (1997) "Development Interventions and Pastoral Dynamics in Southern Ethiopia," in R. Hogg (ed.) *Pastoralists, Ethnicity and the State in Ethiopia*, London: Haan Publishing, pp. 55–80.

—— (1998) "Land Alienation in Borana: Some Land Tenure Issues in a Pastoral Contest in Ethiopia." Paper presented at *OSSREA/University of Bergen Workshop on Human Adaptation in African Drylands*, Khartoum.

Helmsing, A. H. J. (2000) "Decentralisation and Enablement: Issues in the Local Governance Debate," inaugural address, Utrecht University.

Hettne, B. (1996) "Ethnicity and Development: an Elusive Relationship," in D. Dwyer and D. Drakakis-Smith (eds) *Ethnicity and Development: Geographical Perspectives*, London: Wiley & Sons, pp. 15–44.

—— (2002) "Current Trends and Future Options in Development Studies," in V. Desai and R. Potter (eds) *The Companion to Development Studies*, London: Arnold, pp. 7–12.

Hickey, S. and Mohan, G. (eds) 2004 *Participation: From Tyranny to Transformation? Exploring New Approaches to Participation in Development*, London: Zed Books.

Hickox, M. (1999) "Cultural Disease and British Industrial Decline: Weber in Reverse," in T. Skelton and T. Allen (eds) *Culture and Global Change*, London: Routledge, pp. 137–44.

Hoffman, G. (2000) "The Dot.com Pioneers," *Jerusalem Post* (April 23).

Hojman, D. (1999) "Economic Policy and Latin American Culture: is a Virtuous Circle Possible?," *Journal of Latin American Studies*, 31 (1): 67–90.

Hopkins, M. (1999) *The Planetary Bargain*, London: Macmillan.

Hosokawa, S. (2002) "Salsa no tiene Fronteras: Orquesta de la Luz and the Globalization of Popular Music," in L. Waxer (ed.) *Situating Salsa: Global Markets and Local Meanings in Latin Popular Music*, New York and London: Routledge, pp. 289–311.

Hotz-Hart, B. (2000) "Innovation Networks, Regions and Globalization," in G. L. Clark, M. P. Feldman, and M. S. Gertler (eds) *The Oxford Handbook of Economic Geography*, Oxford: Oxford University Press, pp. 432–50.

Hourani, A. and Shehadi, N. (eds) (1993) *Lebanese in the World: A Century of Emigration*, London: I. B. Tauris.

Hudson, R. (1999) "The Learning Economy, the Learning Firm and the Learning Region: a Sympathetic Critique of the Limits to Learning," *European Urban and Regional Studies*, 6 (1): 59–72.

Huntington, S. (1997) *The Clash of Civilizations and the Remaking of the World Order*, New York: Simon & Schuster.

Huqqe, G. (1999) "Borana community," unpublished paper for SOS Sahel Borana Collaborative Forest Management Project, SOS Sahel, Negele Borana.

Hulme, D. and Murphree, M. (eds) (2001) *African Wildlife and Livelihoods: The Promise and Performance of Community Conservation*, Oxford: James Currey.

Hultin, J. (1996) "Perceiving Oromo: 'Galla' in the Great Narrative of Ethiopia," in P. T. W. Baxter, J. Hultin, and A. Triulzi (eds) *Being and Becoming Oromo: Historical and Anthropological Enquiries*, Lawrenceville, NJ: Red Sea Press, pp. 81–91.

Hutt, M. D., Stafford, E. R., Walker, B. A., and Reingen, P. H. (2000) "Case Study: Defining the Social Network of a Strategic Alliance," *Sloan Management Review*, 41 (2): 51–65.

IFAD (International Fund for Agricultural Development) (2005) *Development Project for Indigenous and Afro-Ecuadorian Peoples (PRODEPINE) – Interim Evaluation*, IFAD website, accessed April 1, 2005.

Ignatieff, M. (1996) "Overview: Articles of Faith," *Index on Censorship*, 5: 113.

Iheduru, O. C. (2004) "Black Economic Power and Nation-Building in Post-Apartheid South Africa," *Journal of Modern African Studies*, 42 (1): 1–30.

ILDIS (1993) *Lo Pluri-multi o el Reino de la Diversidad*, La Paz: ILDIS.

Ingold, T. (1986) *The Appropriation of Nature: Essays on Human Ecology and Social Relations*, Manchester: Manchester University Press.

Inter-American Development Bank (2004) *IDB's Social Enterpreneurship Program to Support Indigenous Coffee Growers in Guatemala*, press release, November 19, 2004. Washington, DC: Inter-American Development Bank.

ITDG (Intermediate Technology Development Group) (2005) "Indigenous Ingenious – Ecuador," web page accessed April 26, 2005, www.tve.org.

Iwabuchi, K. (2002a) *Recentering Globalization: Popular Culture and Japanese Transnationalism*, Durham, NC and London: Duke University Press.

—— (2002b) "Nostalgia for a (Different) Asian Modernity: Media Consumption of 'Asia' in Japan," *Positions* 10 (3): 547–73.

Iyer, P. (2000) *The Global Soul: Jet Lag, Shopping Malls and the Search for Home*, New York: Knopf.

Jackson, C. (1997) "Post-Poverty, Gender and Development?" *Institute of Development Studies (IDS) Bulletin*, 28 (3): 145–53.

Jackson, J. (1989) "Is There a Way to Talk About Making Culture Without Making Enemies?," *Dialectical Anthropology*, 14: 127–43.

Jacobs, J. M. (1996) *Edge of Empire: Postcolonialism and the City*, London: Routledge.

James, A. (2003) "Regional Culture, Corporate Strategy and High Tech Innovation: Salt Lake City," unpublished Ph.D. dissertation, University of Cambridge.

—— (2005) "Demystifying the Role of Culture in Innovative Regional Economies," forthcoming in *Regional Studies*, 39 (9).

—— (2006) "Critical Moments in the Production of Rigorous and Relevant Cultural Economic Geographies," forthcoming in *Progress in Human Geography*, 30 (6).

Jameson, F. (1998) *The Cultural Turn*, London, Verso.

—— (1991) *Postmodernism or the Cultural Logic of Late Capitalism:* Durham, NC: Duke University Press.

Japan PHRD (2003) *Japan PHRD Funding Proposal for the Technical Assistance Program: Ecuador, Second Indigenous and Afro-Ecuadorian Peoples Development Project*, accessed via World Bank website, April 1, 2005.

Jensen, S. (2004) "Claiming Community," *Critique of Anthropology*, 24 (2): 179–207.

Johannisson, B., Alexanderson, O., Nowicki, K., and Senneseth, K. (1994) "Beyond Anarchy and Organization: Entrepreneurs in Contextual Networks," *Entrepreneurship and Regional Development*, 6: 329–56.

Johnston, R. J., Gregory, D., Pratt, G., and Watts, M. (eds) (2000) *The Dictionary of Human Geography*, 4th edn, Oxford: Blackwell.

Jolly, S. (2002) *Gender and Cultural Change: Overview Report*, BRIDGE report. Falmer: Institute of Development Studies.

Joseph, M. (2001) *Against the Romance of Community*, Minneapolis: University of Minnesota Press.

Joshi, S. (1998) "Discussion on Microcredit," GREAT network: September 3, 1998.

Kabeer, N. (1994) *Reversed Realities: Gender Hierarchies in Development Thought*, London, Verso.

—— (2000) "Social Exclusion, Poverty and Discrimination: Towards an Analytical Framework," *IDS Bulletin*, 31 (4): 83–97.

Kaplan, C. (1995) "A World Without Boundaries: the Body Shop's Trans/National Geographic," *Social Text*, 43: 45–66.

Katz, C. (2004) *Growing Up Global: Economic Restructuring and Children's Everyday Lives*, London: University of Minnesota Press.

Keane, J. (2003) *Global Civil Society. Civil Society*, Cambridge: Polity Press.

Keeble, D. and Wilkinson, F. (eds) (2000) *High-Technology Clusters, Networking and Collective Learning in Europe*, Aldershot: Ashgate Publishing.

Kelly, J. and Kaplan, M. (2001) *Represented Communities*, Chicago: University of Chicago Press.

Kelly, W. (1991) "Directions in Anthropology of Contemporary Japan," *Annual Review of Anthropology*, 20: 395–431.

Kiely, R. (1999) "The Last Refuge of the Noble Savage? A Critical Assessment of Post-Development Theory," *European Journal of Development Research*, 11 (1): 30–55.

Kirshenblatt-Gimblett, B. (1998) *Destination Culture: Tourism, Museums and Heritage*, Berkeley, CA: University of California Press.

Kitching, G. (1980) *Development and Underdevelopment in Historical Perspective*, London: Methuen.

Kleymeyer, C. (1995) *Cultural Expressions and Grassroots Development*, Quito: Inter-American Foundation and Abya-Yala.

Kliksberg, B. (1999) "Social Capital and Culture: Master Keys to Development," *CEPAL Review*, 69: 83–102.

Kopytoff, I. (ed.) (1987) *The African Frontier: The Reproduction of Traditional African Societies*, Bloomington, IN, Indiana University Press.

Kothari, B. (1997) "Rights to Benefits of Research: Compensating Indigenous People for Their Intellectual Contribution," *Human Organization*, 56 (2): 127–37.

Kothari, U. and Minogue, M. (eds) (2002) *Critical Perspectives in Development Theory and Practice*, Basingstoke, Macmillan.

Kotkin, J. (1993) *Tribes: How Race, Religion, and Identity Determine Success in the New Global Economy*, New York: Random House.

—— (1999) "The New Ethnic Entrepreneurs," *Los Angeles Times* (September 12).

Kotler, K. and Lee, N. (2004) *Corporate Social Responsibility: Doing the Most Good for Your Company and Your Cause*, New York: Wiley.

Krishna, A. (2001) "Moving from the Stock of Social Capital to the Flow of Benefits: the Role of Agency," *World Development*, 29 (6): 925–43.

Krylow, A. (1994) "Ethnic factors in post-Mengistu Ethiopia" in A. Zegeye and S. Pausewang (eds) *Ethiopia in Change: Peasantry, Nationalism and Democracy*, London: British Academic Press, pp. 231–41.

Kumar, A. (ed.) (2003) *World Bank Literature*, Minneapolis: University of Minnesota Press.

Kuran, T. (2004) "Cultural Obstacles to Economic Development: Often Overstated, Usually Transitory," in V. Rao and M. Walton (eds) *Culture and Public Action*, Stanford, CA: Stanford University Press, pp. 115–37.

Kwok Bun, C. (ed.) (2000) *Chinese Business Networks: State, Economy and Culture*, Englewood Cliffs, NJ: Prentice Hall.

Kyei, P. (1999) "Decentralisation and Poverty Alleviation in Rural Ghana," unpublished Ph.D. thesis, Department of Geography, University of Durham.

Kyle, D. (1999) "The Otavalo Trade Diaspora: Social Capital and Transnational Entrepreneurship," *Ethnic and Racial Studies*, 22 (2): 422–46.

Laclau, A. (1996) *Emancipations*, London: Verso.

Laguerre, M. S. (1999) *The Global Ethnopolis: Chinatown, Japantown, and Manila-town in American Society*, New York: St Martin's Press.

Landes, D. (2000) "Culture Makes All the Difference," in L. Harrison and S. Huntington (eds) *Culture Matters*, London: Basic Books, pp. 2–13.

Larner, W. (2003) "Neoliberalism?," *Society and Space*, 21: 509–12.

Larraín, J. (2000) *Identity and Modernity in Latin America*, Cambridge: Polity.

Larreamendy, P. (2002) "Indigenous Networks: Politics and Development Interconnectivity among the Shuar in Ecuador," unpublished Ph.D. dissertation, University of Cambridge.

—— (2003) "Indigenous Networks: Politics and Development Interconnectivity Among the Shuar in Ecuador," unpublished Ph.D. dissertation, Department of Geography, University of Cambridge.

Larson, A. and Ribot, J. (2004) "Democratic Decentralisation Through a Natural Resource Lens: an Introduction," *The European Journal of Development Research*, 16 (1): 1–25.

Larson, B. and Harris, O. (eds) (1995) *Ethnicity, Markets and Migration in the Andes: at the Crossroads of Anthropology and History*, London: Duke University Press.

Lash, S. and Robertson, R. (eds) (1995) *Global Modernities*, London: Sage.

—— and Urry, J. (1994) *Economies of Signs and Space*, London: Sage.

Latour, B. (1991) *We Have Never Been Modern*, Cambridge, MA: Harvard University Press.

Laurie, N. (2005) "Putting the Messiness Back in: Towards a Geography of Development as Creativity. A Commentary on J. K. Gibson-Graham's 'Surplus Possibilities: Post Development and Community Economies'," *Singapore Journal of Tropical Geography* 26 (1): 32–5.

——, Andolina, R., and Radcliffe, S. A. (2002) "The Excluded 'Indigenous'? The Implications of Multi-Ethnic Policies for Water Reform in Bolivia," in R. Sieder (ed.) (2002) *Multiculturalism in Latin America: Indigenous Rights, Diversity and Democracy*, London: Palgrave-Macmillan, pp. 252–76.

—— (2005) "Ethnodevelopment: Social Movements, Creating Experts and Professionalizing Indigenous Knowledge in Ecuador," *Antipode*, 37 (3): 470–96.

Lawson, C. and Lorenz, E. (1999) "Collective Learning, Tacit Knowledge and Regional Innovative Capacity," *Regional Studies*, 33 (4): 305–17.

Lazaric, N. and Lorenz, E. (1997) "Trust and Organisational Learning During Inter-firm Co-operation," in N. Lazaric and E. Lorenz (eds), *The Economics of Trust and Learning*, London: Elgar, pp. 209–26.

Leach, M. and Mearns, R. (eds) (1996) *The Lie of the Land*, Oxford: James Currey.

——, Mearns, R., and Scoones, I. (1997) "Editorial: Community-Based Sustainable Development: Consensus or Conflict," *IDS Bulletin*, 28 (4): 1–3.

——, Scoones, I., and Wynne, B. (eds) (2005) *Science and Citizens*, London: Zed Books.

Lee, G. (2000) "Not Your Father's Detroit," *Washington Post* (January 9).

Leftwich, A. (ed.) (1996) *Democracy and Development: Theory and Practice*, Cambridge: Polity.

Lehmann, D. (ed.) (1982) *Ecology and Exchange in the Andes*, Cambridge: Cambridge University Press.

—— (1997) "An Opportunity Lost: Escobar's Deconstruction of Development," *Journal of Development Studies*, 33 (4): 568–78.

Lewis, D., Bebbington, A., Batterbury, S., Shar, A., Olson, E., Siddiqi, M. S., and Duvall, S. (2003) "Practice, Power and Meaning: Frameworks for Studying Organizational Culture in Multi-Agency Rural Development Projects," *Journal of International Development*, 15: 541–57.

Lewis, O. (1959) *Five Families: Mexican Case Studies in the Culture of Poverty*, New York: Basic Books.

Lewis, S. (2000) "Sanitising South Africa," *Soundings*, 16: 46.

Li, T. (2005) *Governmentality and Its Limits*, Durham, NC: Duke University Press.

Light, I. and Bonacich, E. (1988) *Immigrant Entrepreneurs: Koreans in Los Angeles*, Berkeley, CA: University of California Press.

—— and Gold, S. J. (1999) *Ethnic Economies*, San Diego, CA: Academic Press.

—— and Karageorgis, S. (1994) "The Ethnic Economy," in N. J. Smelser and R. Swedberg (eds) *The Handbook of Economic Sociology*, Princeton, NJ: Princeton University Press, pp. 647–71.

——, Bernard, R., and Kim, R. (1999) "Immigrant Incorporation in the Garment Industry of Los Angeles," *International Migration Review*, 33, (1): 5–25.

——, Sabagh, G., Bozorgmehr, M., and Der-Martirosian, C. (1993) "Internal Ethnicity in the Ethnic Economy," *Ethnic and Racial Studies*, 16 (4): 581–97.

——, Sabagh, G., Bozorgmehr, M., and Der-Martirosian, C. (1994) "Beyond the Ethnic Enclave Economy," *Social Problems*, 41 (1): 65–80.

Lin, J. (1998) *Reconstructing Chinatown: Ethnic Enclave, Global Change*, Minneapolis: University of Minnesota Press.

Liu, H. (1998) "Old Linkages, New Networks: The Globalization of Overseas Chinese Voluntary Associations and its Implications," *The China Quarterly* 155: 582–609.

Loomis, T. (2000) "Indigenous Populations and Sustainable Development: Building on Indigenous Approaches to Holistic, Self-Determined Development," *World Development*, 28 (5): 893–910.

Lorenz, E. H. (1992) "Trust, Community, and Co-operation: Toward a Theory of Industrial Districts," in M. Storper and A. J. Scott (eds) *Pathways to Industrialisation and Regional Development*, London: Routledge, pp. 195–204.

Lowe, L. (1991) "Heterogeneity, Hybridity, Multiplicity: Marking Asian American Differences," *Diaspora*, 1 (1): 24–44.

—— and Lloyd, D. (eds) (1997) *The Politics of Culture in the Shadow of Capital*, London: Duke University Press.

Ludden, D. (1992) "India's Development Regime," in N. Dirks, (ed.) *Colonialism and Culture*, Ann Arbor, MI: University of Michigan Press, pp. 247–88.

Ludlow, D. H. (1992) *The Encyclopedia of Mormonism: The History, Scripture, Doctrine and Procedures of the Church of Latter-Day Saints*, New York: Macmillan.

Lyon, F. (2000) "Trust, Networks and Norms: the Creation of Social Capital in Agricultural Economies in Ghana," *World Development*, 28 (4): 663–82.

—— (2003) "The Relations of Cooperation in Group Enterprises and Associations in Ghana: Exploration of Issues of Trust and Power," in F. Sforzi (ed.) *Institutions of Local Economic Development*, Aldershot: Ashgate, pp. 75–91.

—— and Porter, G. (2005) "The Social Relations of Economic Life and Networks of Civic Engagement: Social Capital and Targeted Development in West Africa," in R. Cline-Cole and E. Robson (eds) *West African Worlds: Local and Regional Paths through Development, Modernisation and Globalisation*, London: Pearson Education, pp. 191–206.

McDowell, L. M. (1997) *Capital Culture: Gender at Work in the City*, Blackwell: Oxford.

—— (2004) "Work, Workfare, Work/Life Balance, and an Ethic of Care," *Progress in Human Geography*, 28 (2): 145–63.

McEwan, C. (2003) "Taking Culture into Development? Discussant's Comments," paper presented at American Association of Geographers annual conference, New Orleans.

MacKinnon, D., Cumbers, A., and Chapman, K. (2002) "Learning, Innovation and Regional Development: a Critical Appraisal of Recent Debates," *Progress in Human Geography*, 26 (3): 293–311.

McNally, D. (1995) *Against the Market*, London: Verso.

McNeill, D. (1996) "Making Social Capital Work: an Extended Comment on the Concept of Social Capital, Based on Robert Putnam's Book," *Forum for Development Studies*, 2: 417–21.

—————— (2003) "Social Capital and the World Bank," in M. Bøås and D. McNeill (eds) *Global Institutions and Development: Framing the World?* London: Routledge, pp. 108–23.

MacPherson, A. (1992) "Innovation, External Technical Linkages and Small-Firm Commercial Performance: an Empirical Analysis from Western New York," *Entrepreneurship and Regional Development*, 4: 165–83.

Maillat, D. (1995) "Territorial Dynamics, Innovative Milieus and Regional Policy," *Entrepreneurship and Regional Development*, 7: 157–65.

Malecki, E. J. and Oinas, P. (eds) (1999) *Making Connections: Technological Learning and Regional Economic Change*, Aldershot: Ashgate.

Malmberg, A. and Maskell, P. (1997) "Towards an Explanation of Regional Specialisation and Industry Agglomeration," *European Planning Studies*, 5: 25–41.

Malthus, T. (1798) *Essay on the Principal of Population*, London: Oxford University Press.

Mamdani, M. (1996) *Citizen and Subject*, Princeton, NJ: Princeton University Press

Markusen, A. (1994) "Studying Regions by Studying Firms," *Professional Geographer*, 46 (4): 477–90.

—— (1999) "Fuzzy Concepts, Scanty Evidence, Policy Distance: the Case for Rigour and Policy Relevance in Critical Regional Studies," *Regional Studies*, 33 (9): 869–84.

Martin, R. L. and Sunley, P. (2001) "Rethinking the 'Economic' in Economic Geography: Broadening Our Vision or Losing Our Focus?," *Antipode*, 33 (2): 148–61.

Martinussen, J. (1997) *Society, State and Market*, London: Zed Books.

Masden, K. (1997) "The Impact of Ministry of Education Policy on Pluralism in Japanese Education: an Examination of Recent Issues," in D. Y. H. Wu, H. McQueen and Y. Yasushi, (eds) *Emerging Pluralism in Asia and the Pacific*, Hong Kong: Hong Kong Institute of Asia-Pacific Studies/The Chinese University of Hong Kong, pp. 29–63.

Massey, D., Quintas, P., and Wield, D. (1992) *High Tech Fantasies: Science Parks in Society, Science and Space*, London: Routledge.

Matthee, R. P. (2000) *The Politics of Trade in Safavid Iran: Silk for Silver, 1600–1730*, Cambridge: Cambridge University Press.

May, D. L. (2001) "Mormons," in E. A. Eliason (ed.), *Mormons and Mormonism: An Introduction to an American World Religion*, Chicago: University of Illinois Press, pp. 47–75.

Mayer, M. and Rankin, K. N. (2002) "Social Capital and (Community) Development: A North/South Perspective," *Antipode*, 34 (4): 804–8.

Mayoux, L. (2001) "Tackling the Down Side: Social Capital, Women's Empowerment and Microfinance in Cameroon," *Development and Change*, 32: 435–64.

Medina, L. K. (2003) "Commoditizing Culture: Tourism and Maya Identity," *Annals of Tourism Research*, 30 (2): 353–68.

Meisch, L. (2002) *Andean Entrepreneurs*, Austin, TX: University of Texas Press.

Melendez, C. S. (2000) "Vietnamese, Mexican Émigrés Find Common Ground at Checkout Stand," *San Jose Mercury News* (January 14).

Mertes, T. (2003) *Movement of Movements*, London: Verso.

Midgley, J. (1995) *Social Development: The Developmental Perspective in Social Welfare*, London: Sage.

Milkman, R. (ed.) (2000) *Organizing Immigrants: The Challenge for Unions in Contemporary California*, Ithaca, NY: Cornell University Press.

Minghuan, L. (2000) *"We Need Two Worlds": Chinese Immigrant Associations in a Western Society*, Amsterdam: Amsterdam University Press.

Mishra, A., Vasavada, S., and Bates, C. (in press) "How Many Committees Do I Belong To?" in R. Jeffery and N. Sundar (eds) *Organising Sustainability: NGOs and Joint Forest Management in India*, London: Sage.

Misztal, B. A. (1996) *Trust*, Cambridge: Polity Press.

Mitchell, D. (2000) *Cultural Geography: a Critical Introduction*, Blackwell: Oxford.

Mitchell, T. (2000) "The Stage of Modernity" in T. Mitchell (ed.) *Questions of Modernity*, Minneapolis: University of Minnesota Press, pp. 1–34.

—— (2002) *Rule of Experts: Egypt, Techno-Politics, Modernity*, Berkeley, CA: University of California Press.

Mitsui, T. (1993) "The Reception of the Music of American Southern Whites in Japan," in N. V. Rosenberg (ed.) *Transforming Tradition: Folk Music Revivals Examined*, Chicago: University of Illinois Press, pp. 275–93.

MNghi Ha, K. (2000) "Ethnizität, Differenz und Hybridität in der Migration: eine Postkoloniale Perspektive," *Prokla, Zeitschrift für Kritische Sozialwissenschaft*, 29 (3).

Mohan, G. (2002) "The Disappointments of Civil Society: The Politics of NGO Intervention in Northern Ghana," *Political Geography*, 21 (1): 125–54.

Mohanty, C. T., Russo, A., and Torres, A. (eds) (1991) *Third World Women and the Politics of Feminism*, Bloomington, IN: University of Indiana Press.

Molyneux, M. (2002) "Gender and the Silences of Social Capital: Lessons from Latin America," *Development and Change*, 33 (2): 167–88.

—— and Lazar, S. (2003) *Doing the Rights Thing: Rights Based Development and Latin American NGOs*, London: ITDG Publications.

Momsen, J. H. (2001) "Backlash: or How to Snatch Failure from the Jaws of Success in Gender and Development," *Progress in Development Studies*, 1 (1): 51–6.

Morrow, S. and Vokwana, N. (2001) "'Shaping in Dull, Dead Earth Their Dreams of Riches and Beauty': Clay Modelling at e-Hala and Hogsback in the Eastern Cape, South Africa," *Journal of Southern African Studies*, 27 (1): 137–61.

Mosse, D. (2001) "'People's Knowledge': Participation and Patronage: Operations and Representations in Rural Development," in B. Cooke and U. Kothari (eds) *Participation: The New Tyranny?*, London: Zed Books, pp. 16–35.

Murphy-Shigematsu, S. (2000) "Identities of Multiethnic People in Japan," in M. Douglass and G. S. Roberts, (eds) *Japan and Global Migration: Foreign Workers and the Advent of a Multicultural Society*, London: Routledge, pp. 196–216.

Narayan, D. and Pritchett, F. (1997) *Cents and Sociability: Household Income and Social Capital in Rural Tanzania*, Policy Research Working Paper no. 1796, Washington, DC: World Bank.

Nederveen Pieterse, J. (1995) "Globalization as Hybridization," in M. Featherstone, S. Lash, and R. Robertson (eds) *Global Modernities*, London: Sage, pp. 45–68.

—— (1997a) "Deconstructing/Reconstructing Ethnicity," *Nations and Nationalism*, 3 (3): 1–31.

—— (1997b) "Traveling Islam: Mosques without Minarets," in A. Öncü and P. Weyland (eds) *Space, Culture and Power*, London: Zed Books, pp. 177–200.

—— (1998) "My Paradigm or Yours? Alternative Development, Post-Development, Reflexive Development," *Development and Change*, 29: 343–73.

—— (2000) "Globalization and Human Integration: We Are All Migrants," *Futures*, 32 (5): 385–98.

—— (2001a) *Development Theory: Deconstructions/Reconstructions*, London: Sage.

—— (2001b) "Hybridity, So What? The Anti-hybridity Backlash and the Riddles of Recognition," *Theory, Culture & Society*, 18 (2/3): 1–27.

—— (2002) "Ethnicities and Multiculturalisms: Drawing Boundaries," in S. May, T. Modood, and J. Squires (eds) *Ethnicity, Nationalism and Minority Rights*, Cambridge: Cambridge University Press, pp. 27–49.

Nelson, C. and Grossberg, G. (eds) (1988) *Marxism and the Interpretation of Culture*, Urbana, IL: University of Illinois Press.

Nelson, N. and Wright, S. (eds) 1995 *Power and Participatory Development: Theory and Practice*, London: IT Press.

Newton, K. (1999) "Social Capital and Democracy in Modern Europe," in J. W. van Deth, M. Maraffi, K. Newton, and P. F. Whiteley (eds) *Social Capital and European Democracy*, London: Routledge, pp. 3–24.

Niethammer, L. (2003) "The Infancy of Tarzan," *New Left Review*, 19: 79–92.

Noguchi, Y. (1999) "Matching Faith and Finance: Alternatives to Loans Cater to Area Muslims," *The Washington Post* (Oct 28).

Norman, J. (2000) "Diverse Industries Covered in Latina Business Survey," *Orange County Register*, (Sept. 26).

North, L. and Cameron, J. (2000) "Grassroots Rural Development Strategies: Ecuador in Comparative Perspective," *World Development*, 28 (10): 1751–66.

Norton, R. D. (2001) *Creating the New Economy: The Entrepreneur and US Resurgence*, Cheltenham: Edward Elgar.

Nuttall, S. and Coetzee, C. (1998) "Introduction" in *Negotiating the Past: The Making of Memory in South Africa*, Oxford: Oxford University Press, pp. 1–15.

Obbo, C. (1979) "Village Strangers in Buganda Society," in W. A. Shack and E. P. Skinner (eds) *Strangers in African Societies*, Berkeley, CA: University of California Press, pp. 227–42.

Oc, T. and Tiesdell, S. (1999) "Supporting Ethnic Minority Business: a Review of Business Support for Ethnic Minorities in City Challenge Areas," *Urban Studies*, 36 (10): 1723–46.

Ocampo Thomason, P. (in press) "Mangroves, People and Cockles: the Impacts of Shrimp Farming Industries on Mangrove Communities in the Esmeraldas Province, Ecuador," in *Environment and Livelihood in the Tropical Coastal Zones: Managing Agriculture-Fishery-Aquaculture Conflicts*, Wallingford: CABI Publishing.

OECD (1999) *Managing National Innovation Systems*, Paris: OECD.

Ohmae, K. (1992) *The Borderless World: Power and Strategy in the Global Marketplace*, London: Collins.

Oinas, O. (1995) "Discussion of 'Regional Advantage: Culture and Competition in Silicon Valley and Route 128' by AnnaLee Saxenian," *Economic Geography*, 71: 202–4.

Oinas, P. and Malecki, E. J. (2002) "The Evolution of Technologies in Time and Space: from National and Regional to Spatial Innovation Systems," *International Regional Science Review*, 25 (1): 102–31.

Okonta I. and Douglas, O. (2001) *Where Vultures Feast*, San Francisco, CA: Sierra Club.

Omi, M. and Winant, H. (1994) *Racial Formation in the United States: from the 1960s to the 1990s*, New York: Routledge.

Ong, A. (1999) *Flexible Citizenship: The Cultural Logics of Transnationality*, Durham, NC and London: Duke University Press.

Ooka, E. and Wellman, B. (2006) "Does Social Capital Pay off More within or between Ethnic Groups? Analyzing Job Searchers in Five Toronto Ethnic Groups," in E. Fong (ed.) *Inside the Mosaic*, Toronto, ON: Toronto University Press (forthcoming).

Osborne, T. and Rose, N. (eds) (1998) "Society and the Life Sciences," *Economy and Society*, 27 (2–3).

O'Reilly, C.A. (1989) "Corporations, Culture, and Commitment: Motivation and Social Control in Organizations, " *California Management Review*, Summer, 9–25.

Ostling, R. and Ostling, J. (1999) *Mormon America: The Power and the Promise*, San Francisco, CA: Harper Collins.

Ostrom, E. (1990) *Governing the Commons: The Evolution of Institutions for Collective Action*, Cambridge: Cambridge University Press.

Ouchi, W. G. (1981) *Theory Z: How American Business Can Meet the Japanese Challenge*, Reading, MA: Addison-Wesley.

Owusu, T. Y. (2000) "The Role of Ghanaian Immigrant Associations in Toronto, Canada," *International Migration Review*, 34 (4): 1155–81.

Oxfam America (2004) "Preserving Bolivia's Ancient Culture," Oxfam America website, accessed December 15, 2004.

Özcan, V. and Seifert, W. (2000) "Self-employment among Immigrants in Germany: Exclusion or Path to Integration?," *Soziale Welt*, 51 (3): 289–302.

Pacari, N. (1996) "Taking on the Neoliberal Agenda," *NACLA Report on the Americas*, 29 (5): 23–32.

Parekh, B. (1997) "South Asians in Britain," *History Today*, 47 (9): 65–8.

Parry, B. (2002) "Cultures of Knowledge: Investigating Intellectual Property Rights and Relations in the Pacific," *Antipode*, 34 (4): 670–706.

Partridge, W and Uquillas, E. (1996) *Including the Excluded: Ethnodevelopment in Latin America*, unpublished document, Latin American and Caribbean Technical Department Environment Unit, Washington, DC: World Bank.

Pascale, R. T. and Athos, A. G. (1982) *The Art of Japanese Management: Applications for American Executives*, New York: Warner Books.

Peck, J. (1999) "Grey Geography," *Transactions of the Institute of British Geographers*, 24: 131–5.

Perreault, T. and Martin, P. (eds) (2005) "Geographies of Neo-liberalism in Latin America," *Environment and Planning A*, 37 (2).

Piguet, E. (1999) *Les Migrations Créatrices: Étude de l'Entreprenariat des Étrangers en Suisse*, Paris: L'Harmattan.

Pirttijarvi, J. (1999) *Indigenous Peoples and Development in Latin America*, report commissioned by Department of International Development Co-operation, Ministry of Foreign Affairs, Finland.

Platt, T. (1982) "The Role of the Andean Ayllu in the Reproduction of the Petty Commodity Regime in Northern Potosí," in D. Lehmann (ed.) *Ecology and Exchange in the Andes*, Cambridge: Cambridge University Press, pp. 27–69.

Platteau, J-P. (1994) "Beyond the Market Stage Where Real Societies Exist: Parts I and II," *Journal of Development Studies*, 30: 533–77, 753–817.

—— (2000) *Institutions, Social Norms and Economic Development*, London, Harwood.

—— (2004) "Participatory Development: Where Culture Creeps" in V. Rao and M. Walton (eds) *Culture and Public Action: A Cross-Disciplinary Dialogue on Development Policy*, Stanford, CA: Stanford University Press, pp. 210–33.

Polanyi, K. (1944) *The Great Transformation*, Boston, MA: Beacon Press.

—— (1945) *The Origins of Our Time*, Boston, MA: Beacon Press.

Poll, R. D. (2001) "Utah and the Mormons: a Symbiotic Relationship," in E. A. Eliason (ed.) *Mormons and Mormonism: An Introduction to an American World Religion*, Urbana, IL: University of Illinois Press, pp. 164–79.

Porter, D., Allen, B., and Thompson, G. (1991) *Development in Practice: Paved with Good Intentions*, London: Routledge.

Porter, G. (2002a) "Living in a Walking World: Rural Mobility and Social Equity Issues in Sub-Saharan Africa," *World Development*, 30 (2): 285–300.

—— (2002b) "Improving Mobility and Access for the Off-Road Rural Poor through Intermediate Means of Transport," *World Transport Policy and Practice*, 8 (4): 6–19.

—— (2003) "NGOs and Poverty Reduction in a Globalizing World: Perspectives from Ghana," *Progress in Development Studies*, 3 (2): 131–45.

Porter, M. E. (1990) *The Competitive Advantage of Nations*, London: Macmillan.

—— (1996) "Competitive Advantage, Agglomeration Economies, and Regional Policy," *International Regional Science Review*, 1 (1/2): 85–94.

—— (1998) *On Competition*, Boston, MA: Harvard University Press.

Portes, A. (1987) "The Social Origins of the Cuban Enclave Economy of Miami," *Sociological Perspectives*, 30 (4): 340–72.

—— (1994) "The Informal Economy and its Paradoxes," in N. J. Smelser and R. Swedberg (eds) *The Handbook of Economic Sociology*, Princeton, NJ: Princeton University Press, pp. 426–50.

—— (1995) *The Economic Sociology of Immigration*, New York: Russell Sage Foundation.

—— (1996) "Transnational Communities: Their Emergence and Significance in the Contemporary World-system," in R. P. Korzeniewicz and W. C. Smith (eds) *Latin America in the World Economy*, Westport, CT: Greenwood Press, pp. 151–68.

—— (2000) "Globalization from Below: The Rise of Transnational Communities," in D. Kalb, M. van der Land, R. Staring, B. van Steenbergen, and N. Wilterdink (eds) *The Ends of Globalization: Bringing Society Back In*, Lanham, MD: Rowman & Littlefield, pp. 253–70.

—— and Landolt, P. (1996) "The Downside of Social Capital," *The American Prospect*, 26: 18–21, 94.

—— and Landolt, P. (2000) "Social Capital: Promise and Pitfalls of Its Role in Development," *Journal of Latin American Studies*, 32 (2): 529–47.

—— and Sensenbrenner, J. (1993) "Embeddedness and Immigration: Notes on the Social Determinants of Economic Action," *American Journal of Sociology*, 98 (6): 1320–50.

Povey, M. (1998) *A History of the Modern Fact*, Chicago: University of Chicago Press.

Power, M. (2003) *Rethinking Development Geographies*, London: Routledge.

Prabhu, A. (2003) "Mariama Bâ's *So Long a Letter:* Women, Culture and Development from a Francophone/Postcolonial Perspective," in K-K. Bhavnani, J. Foran, and P. Kurian (eds) *Feminist Futures, Re-imagining Women, Culture and Development*, London: Zed Books, pp. 239–55.

Prah, K. (2001) "Culture: The Missing Link in Development Planning in Africa," *Présence Africaine*, 163/4: 90–102.

Prahalad, C. and Porter, M. (2003) *Harvard Business Review on Corporate Responsibility*, Cambridge, MA: Harvard Business School Press.

Putnam, R. (1993) *Making Democracy Work: Civil Traditions in Modern Italy*, Princeton, NJ: Princeton University Press.

—— (1995) "Bowling Alone: America's Declining Social Capital," *Journal of Democracy*, 6 (1): 65–78.

—— (2000) *Bowling Alone: The Collapse and Revival of American Community*, New York: Simon & Schuster.

Quirk, J. (2003) "Bureaucratic Culture and the New Imperialist Agenda in India," unpublished Ph.D. thesis, Department of Geography, University of Durham.

Radcliffe, S. A. (2002) "Indigenous Women, Rights and the Nation-State in the Andes," in N. Craske and M. Molyneux (eds) *Gender and the Politics of Rights and Democracy in Latin America*, London: Palgrave, pp. 149–72.

—— (2004) "Development, Civil Society and Inequality I: Social Capital Is (Almost) Dead?" *Progress in Human Geography*, 28 (4): 517–27.

—— (2005) "Development and Geography: Towards a Postcolonial Development Geography?," *Progress in Human Geography*, 29 (3): 291–8.

—— and Laurie, N. (2006) "Culture and Development: Taking Culture Seriously in Development for Andean Indigenous People" *Environment and Planning D: Society and Space*, 24 (2): 1–18.

——, Laurie, N., and Andolina, R. (2004) "The Transnationalization of Gender and Re-Imagining Andean Indigenous Development," *Signs: Journal of Women in Culture and Society*, 29 (2): 387–416.

——, Laurie, N., and Andolina, R. (forthcoming) "Development with Identity: Social Capital and Culture" in R. Andolina, N. Laurie, and S. A. Radcliffe, *Multiethnic Transnationalism: Indigenous Development in the Andes*, Durham, NC and London: Duke University Press.

Rahnema M. and Bawtree, V. (eds) (1997) *The Post Development Reader*, London: Pluto.

Ramphele, M. (1996) *A Life* (2nd edn), Cape Town: David Philip.

Rankin, K. N. (2003) "Anthropologies and Geographies of Globalization," *Progress in Human Geography*, 27 (6): 708–34.

Rao, V. and Walton, M. (2004) "Culture and Public Action: Relationality, Equality of Agency and Development," in V. Rao and M. Walton (eds) *Culture and Public Action: A Cross-Disciplinary Dialogue on Development Policy*, Stanford, CA: Stanford University Press, pp. 3–36.

Rath, J. (1999) "The Informal Economy as Bastard Sphere of Social Integration: The Case of Amsterdam," in E. Eichenhofer (ed.) *Migration und Illegalität*, Osnabrück, Rasch Verlag, pp. 117–36.

—— (ed.) (2000) *Immigrant Businesses: The Economic, Political and Social Environment*, London and New York: Macmillan/St Martin's Press.

Ray, L. and Sayer, A. (eds) (1999) *Culture and Economy After the Cultural Turn*, London: Sage.

RETORT (2005) *Afflicted Powers*, London: Verso.

Richter, K. (1999) "Secret of Success for Many Turks in Germany Lies in Start-ups," *Wall Street Journal* (July 13).

Rigg, J. (1997) *Southeast Asia. The Human Landscape of Modernization and Development*, London: Routledge.

Rist, G. (1997) *The History of Development*, London: Zed Books.

Robinson, L. J., Schmid, A. A., and Siles, M. E. (2002) "Is Social Capital Really Capital?," *Review of Social Economy*, 60 (1): 1–21.

Roitman, J. (2005) *Fiscal Disobedience*, Princeton, NJ: Princeton University Press.

Romney, L. (1999) "Minority-owned Firms Tend to Hire within Their Own Ethnic Group," *Los Angeles Times* (September 18).

Rosaldo, R. (1989) *Culture and Truth: The Remaking of Social Analysis*, Boston: Beacon Press.

Rose, N. (1999) *Powers of Freedom*, Cambridge: Cambridge University Press.

Rose, R., Mishler, W., and Haerpfer, C. (1997) "Social Capital in Civic and Stressful Societies," *Studies in Comparative International Development*, 32 (3): 85–111.

Rowe, W. and Schelling, V. (1991) *Memory and Modernity in Latin America*, London: Verso.

Sabel, C. (1992) "Studied Trust: Building New Forms of Co-operation in a Volatile Economy," in F. Pyke and W. Sengenberger (eds) *Industrial*

Districts and Local Economic Regeneration, Geneva: Institute for Labour Studies, pp. 215–51.

Sachs, W. (1992) *The Development Dictionary*, London: Zed Books.

Said, E. (1978) *Orientalism*, New York: Vintage.

Sassen, S. (1998) *Globalization and its Discontents*, New York: The New Press.

Satha-Anand, C. (1998) "Spiritualising Real Estate, Commoditising Real Estate: the Muslim Minority in Thailand," in J. A. Camilleri and C. Muzaffar (eds) *Globalization: The Perspectives and Experiences of the Religious Traditions of Asia Pacific*, Petaling Jaya: International Movement for a Just World, pp. 135–46.

Saunders, M. (1999) "Sweeteners for Business Migrants," *The Australian* (November 24).

Savigliano, M. (1995) *Tango and the Political Economy of Passion*, Boulder, CO: Westview Press.

Saxenian A. (1989) "The Cheshire Cat's Grin: Innovation, Regional Development and the Cambridge Case," *Economy and Society*, 18 (4): 448–77.

—— (1994) *Regional Advantage: Culture and Competition in Silicon Valley and Route 128*, Cambridge, MA: Harvard University Press.

Sayer, A. (1997) "The Dialectic of Culture and Economy," in R. Lee and J. Wills (eds) *Geographies of Economies*, London: Arnold, pp. 16–27.

—— and Walker, R. (1992) *The New Social Economy: Reworking the Division of Labour*, Oxford: Blackwell.

Schech, S. and Haggis, J. (2000) *Culture and Development: a Critical Introduction*, Blackwell: Oxford.

Schein, E. H. (1992) *Organizational Culture and Leadership* (2nd edn, 1997), San Francisco, CA: Jossey-Bass.

Schieffelin, E. (1998) "Problematizing Performance," in F. Hughes-Freeland (ed.) *Ritual, Performance, Media*, London and New York: Routledge, pp. 194–207.

Schlee, G. (1992) "Ritual Topography and Ecological Use: The Gabbra of the Kenyan/Ethiopian Borderlands," in E. Croll and D. Parkin (eds) *Bush Base: Forest Farm: Culture, Environment and Development*, London: Routledge, pp. 110–30.

—— (1994) "Ethnicity Emblems, Diacritical Features, Identity Markers: Some East African Examples," in D. Brokensha (ed.) *A River of Blessings*, New York: Syracuse University Press, pp. 129–43.

Schmidt, D. (2000) "Unternehmertum und Ethnizität: ein Seltsames Paar," *Prokla, Zeitschrift für Kritische Sozialwissenschaft*, 29 (3): 120.

Schoenberger, E. (1994) "Corporate Strategy and Corporate Strategists: Power, Identity, and Knowledge Within the Firm," *Environment and Planning A*, 26 (3): 435–51.

—— (1997) *The Cultural Crisis of the Firm*, Oxford: Blackwell.

—— (2002) "Creating the Corporate World," in E. Sheppard and T. Barnes, (eds) *A Companion to Economic Geography*, London, Blackwell, pp. 377–91.

Schofield, B. (2002) "Partners in Power: Governing the Self-Sustaining Community." *Sociology* 36: 663.

Scoones, I. (ed.) (1995) *Living with Uncertainty: New Directions for Pastoral Development in Africa*, London: IT Press.

Scott, A. J. (1986) "Industrial Organization and Location: Division of Labour, the Firm, and Social Process," *Economic Geography*, 62: 215–31.

—— (1988) *New Industrial Spaces: Flexible Production, Organization and Regional Development in North America and Western Europe*, London: Pion.

—— (2000) "Economic Geography: The Great Half Century," in G. L. Clark, M. P. Feldman, and M. S. Gertler (eds) *Handbook of Economic Geography*, Oxford: Oxford University Press, pp. 18–44.

Scott, J. (1998) *Seeing Like A State*, New Haven: Yale University Press.

Sen, A. (1999) *Development as Freedom*, New York: Knopf Press.

—— (2004) "How Does Culture Matter?" in V. Rao and M. Walton (eds) *Culture and Public Action*, Stanford: Stanford University Press, pp. 37–58.

Sewell, W. (1999) "The Concept(s) of Culture," in V. Bonnell and L. Hunt (eds) *Beyond the Cultural Turn*, Berkeley, CA: University of California Press, pp. 35–61.

Shack, W. A. (1979) "Introduction," in W. A. Shack and E. P. Skinner (eds) *Strangers in African Societies*, Berkeley, CA: University of California Press, pp. 1–17.

Shepherd, G. and Shepherd, G. (1984) *A Kingdom Transformed: Themes in the Development of Mormonism*, Salt Lake City, UT: University of Utah Press.

Shongolo, A. A. (1996) "The Poetics of Nationalism," in P. T. W. Baxter, J. Hultin, and A. Triulzi (eds) *Being and Becoming Oromo: Historical and Anthropological Enquiries*, Lawrenceville, NJ: Red Sea Press, pp. 265–90.

Shore, C. and Wright, S. (eds) (1997) *Anthropology of Policy*, London: Routledge.

Shupe, A. (1992) *Wealth and Power in American Zion*, Lampeter: Mellen Press.

Sieder, R. (ed.) (2002) *Multiculturalism in Latin America: Indigenous Rights, Diversity and Democracy*, London: Palgrave-Macmillan.

Sikkink, L. (2001) "Traditional Medicines in the Market Place: Identity and Ethnicity Among Female Vendors," in L. Seligmann (ed.) *Women Traders in Cross-Cultural Perspective: Meditating Identities, Marketing Wares*, Stanford: Stanford University Press, pp. 209–25.

Silver, B. and Arrighi, G. (2003) "Polanyi's 'Double Movement': The *Belles Epoches* of British and US Hegemony Compared," *Politics and Society*, 31: 325–55.

Simone, A. (2005) *For the City Yet to Come*, Durham, NC: Duke University Press.

Smart, A. (1993) "Gifts, Bribes and Guanxi: A Reconsideration of Bourdieu's Social Capital," *Cultural Anthropology*, 8 (3): 388–408.

Smith, C. (2003) "Corporate Social Responsibility: Whether and How?," *California Management Review*, 45 (1): 52–76.

Smith, N. (1984) *Uneven Development*, Blackwell: Oxford.

Spear, T. and Waller, R. (1993) *Being Maasai: Ethnicity and Identity in East Africa*, Oxford: James Currey.

Spender, J. C. (1996) "Making Knowledge the Basis for a Dynamic Theory of the Firm," *Strategic Management Journal*, 17: 45–62.

Staber, U. (2003) "Social Capital or Strong Culture?" *Human Resource Development International*, 6 (3): 413–20.

Stanley, E. (2001) "Evaluating the Truth and Reconciliation Commission," *Journal of Modern African Studies*, 39 (3): 525–46.

Stark, R. (1996) "Religion as Context: Hellfire and Delinquency One More Time," *Sociology of Religion*, 57 (2): 163–73.

Starn, O. (1991) "Missing the Revolution: Anthropologists and the War in Peru," *Cultural Anthropology*, 6 (1): 63–91.

Stavenhagen, R. (1996) "Indigenous Rights: Some Conceptual Problems" in E. Jelin and E. Hershberg (eds) *Constructing Democracy: Human Rights, Citizenship and Society in Latin America*, Boulder, CO: Westview, pp. 141–59.

Stearns, P. W. (2001) *Cultures in Motion: Mapping Key Contacts and Their Imprints in World History*, New Haven, CT: Yale University Press.

Stedman Jones, G. (2004) *The End of Poverty*, London: Profile.

Stephen, L. (1991) "Culture as a Resource: Four Cases of Self-Managed Indigenous Craft Production in Latin America," *Economic Development and Cultural Change*, 39: 101–30.

Stirrat, R. (2000) "Cultures of Consultancy," *Critique of Anthropology*, 20 (1): 31–46.

Storper, M. (1995) "The Resurgence of Regional Economies Ten Years Later: the Region As a Nexus of Untraded Interdependencies," *European Urban and Regional Studies*, 2: 191–221.

—— (1997) "Regional Economies As Relational Assets," in R. Lee and J. Wills (eds) *Geographies of Economies*, London: Arnold, pp. 248–58.

Strobele-Gregor, J. (1996) "Culture and Political Practice of the Aymara and Quechua in Bolivia: Autonomous Forms of Modernity in the Andes," *Latin American Perspectives*, issue 89, 23 (2): 72–90.

Sullivan, R. (ed.) (2003) *Business and Human Rights*, Sheffield: Greenleaf.

Sunley, P. (2000) "Urban and Regional Growth," in E. Sheppard and T. J. Barnes (eds), *A Companion To Economic Geography*, Oxford: Blackwell, pp. 187–201.

Suzuki, K., Kim, S. H., and Bae, Z. T. (2002) "Entrepreneurship in Japan and Silicon Valley: a Comparative Study." *Technovation*, 22 (10): 595–606.

Swann, G. M. P., Prevezer, M., and Stout, D. (eds) (1998) *The Dynamics of Industrial Clustering: International Comparisons in Computing and Biotechnology*, Oxford: Oxford University Press.

Tambiah, S. J. (2000) "Transnational Movements, Diaspora, and Multiple Modernities," *Daedalus*, 129 (1): 163–94.

Taylor, C. (2004) *Modern Social Imaginaries*, Durham, NC: Duke University Press.

Taylor, J. G. (1986) *The Social World of Batavia: European and Eurasian Dutch in Asia*, Madison, WI: University of Wisconsin Press.

Temple, J. and Johnson, P. (1996) "Social Capability and Economic Development," unpublished paper, Nuffield College, Oxford.

Tene, C., Tobar, G., and Bolaños, D. (2004) *Programas de Microcrédito y Capital Social entre Mujeres Indígenas*, World Bank Unit for Environmentally and Socially Sustainable Development, Working Paper 18. Quito: World Bank.

Tennant, C. (1994) "Indigenous Peoples, International Institutions and the International Legal Literature from 1945–1993," *Human Rights Quarterly*, 16 (1): 1–57.

Toth, M. A. (1974) "Mormon Society and the United Order: An Analysis of Success and Failure," in G. M. Vernon (ed.) *Research on Mormonism*, Salt Lake City, UT: The Association for the Study of Religion, pp. 174–91.

Townsend, J. with Arrevillaga, V., Bain, J., and Cancino, S. *et al.* (1995) *Women's Voices from the Rainforest*, London: Routledge.

Trouillot, M-R. (2002) "*Adieu* Culture: A New Duty Arises," In R. G. Fox and B. King (eds) *Anthropology Beyond Culture*, London: Routledge, pp. 37–60.

Tseng, Y.-F. (1994) "Chinese Ethnic Economy: San Gabriel Valley, Los Angeles County," *Journal of Urban Affairs*, 16 (2): 169–89.

—— (2000) "The Mobility of Entrepreneurs and Capital: Taiwanese Capital-linked Migration," *International Migration*, 38 (2): 143–68.

Tsing, A. (2005) *Fictions*, Princeton, NJ: Princeton University Press.

Uquillas, J. (2002) "Fortalecimiento de la Capacidad de Autogestión de las Organizaciones Indígenas en el Ecuador: el caso de PRODEPINE" [Strengthening of the Capacity for Self-Management of Ecuadorian Indigenous Organizations: The Case of PRODEPINE]. Paper presented at the 1st Latin American Studies Association Conference on Ecuadorian Studies, Quito, Ecuador.

US Department of Labor (1996) *Employee Benefits in Small Private Industrial Establishments*, Washington, DC: US Bureau of Labor Statistics.

US National Center on the Educational Quality of the Workforce (1995) *The Other Shoe: Education's Contribution to the Productivity of Establishments*, National Employer Survey, EQW Catalogue Number RE02.

UN (United Nations) Preamble (2003) *Preamble to Conference in Stockholm*. Website.

UNDP (United Nations Development Program) (2004) *Human Development Report: Cultural Liberty in Today's Diverse World*, Washington, DC: UNDP.

UNESCO (2003) "Culture and Development" on portal.unesco.org/culture/en/ev (accessed on 29 September 2003).

UNHCR (1997) *The State of the World's Refugees: A Humanitarian Agenda*, Oxford: Oxford University Press.

UN Report (1998) *UN report on Stockholm Intergovernmental Conference on Culture and Development*, New York: United Nations.

Utah Department of Workforce Services (2001) *Annual Report of Labor Market Information, (2000)*, Salt Lake City, UT: Utah Department of Workforce Services.

Valdivia, G. (2005) "On Indigeneity, Change and Representation in the North-eastern Ecuadorian Amazon" *Environment and Planning A*, 37: 285–303.

van Cotthem, C. V. (1999) "More than Market: the Field of Business in *Intercultural Space*," Research Paper. The Hague: Institute of Social Studies.

van Deth, J. W., Maraffi, M., Newton, K., and Whiteley, P. F. (eds) (1999) *Social Capital and European Democracy*, London: Routledge.

Van Nieuwkoop, M. and Uquillas, J. (2000) *Defining Ethnodevelopment in Operational Terms: Lessons from the Ecuadorian Indigenous and Afro-Ecuadorian Peoples Development Project*, Latin American and Caribbean Region, Sustainable Development, Working Paper no.6, Washington DC: World Bank.

van Schaik, L. (1994) "The Netherlands Government's Policy on Indigenous People" in W. Assies and A. Hoekema (eds) *Indigenous Peoples' Experiences with Self-Government*, IWGIA Seminar paper, 79 (1): 203–5. Copenhagen: IWGIA.

Varadarajan, T. (1999) "A Patel Motel Cartel," *New York Times Magazine* (July 4): 36–9.

Vernon, G. M. (1980) *Mormonism: A Sociological Perspective*, Salt Lake City, UT: University of Utah.

Vinding, D. (ed.) (1994) *Indigenous Women: The Right to a Voice*, IWGIA Document 88, Copenhagen: IWGIA.

Visweswaran, K. (1998) "Race and the Culture of Anthropology" *American Anthropologist*, 100 (1): 70–83.

Wade, P. (1999) "Working Culture: Making Cultural Identities in Cali, Colombia," *Current Anthropology*, 40 (4): 449–71.

Walder, D. (2000) "The Necessity of Error: Memory and Representation in the New Literatures," in S. Nasta (ed.) *Reading the "New" Literatures in a Postcolonial Era*, Cambridge: D. S. Brewer, pp. 149–70.

Waldinger, R. (1995) "The 'Other Side' of Embeddedness: A Case-study of the Interplay of Economy and Ethnicity," *Ethnic and Racial Studies*, 18 (3): 555–80.

——, Aldrich, H., and Ward, R. (1990) *Ethnic Entrepreneurs: Immigrant Business in Industrial Societies*, Newbury Park, CA: Sage.

Watson, E. E. (2003) "Examining the Potential of Indigenous Institutions for Development: A Perspective from Borana, Ethiopia," *Development and Change*, 34 (2): 287–309.

Watts, M. (2000) "Poverty and the Politics of Alternatives at the Turn of the Century" in J. Nederveen Pieterse (ed.) *Global Futures*, London: Zed Books, pp. 133–47.

—— (2002) "Alternative Modern – Development As Cultural Geography," in K. Anderson, M. Domosh, S. Pile, and N.Thrift (eds) *Handbook of Cultural Geography*, London: Sage, pp. 433–53.

—— (2003) "Development and Governmentality," *Singapore Journal of Tropical Geography*, 24 (1): 6–34.

—— (2005) "Righteous Oil," *Annual Review of Resources and the Environment*, 9 (1): 9–35.

Weber, M. (1978) *Economy and Society: An Outline of Interpretative Sociology, Volume 1*, Berkeley, CA: University of California Press.

Weinberger, K. and Jutting, J. P. (2001) "Women's Participation in Local Organisations: Conditions and Constraints," *World Development*, 29 (8): 1391–404.

Werbner, P. (2000) "What Colour 'Success'? Distorting Value in Studies of Ethnic Entrepreneurship," in H. Vermeulen and J. Perlmann (eds) *Immigrants, Schooling and Social Mobility: Does Culture Make a Difference?*, London and New York: Macmillan/St Martin's Press, pp. 34–60.

—— (2001) "Metaphors of Spatiality and Networks in the Plural City: a Critique of the Ethnic Enclave Economy Debate," *Sociology*, 35 (3): 671–93.

Werlin, H. (2003) "Poor Nations: Rich Nations: A Theory of Governance," *Public Administration Review*, 63 (3): 329–39.

Wertheim, W. F. (1964) "The Trading Minorities in Southeast Asia," in *East–West Parallels*, The Hague: Van Hoeve, pp. 39–82.

—— (1978) *Indonesië van Vorstenrijk tot Neo-kolonie*, Amsterdam: Boom.

Westwood, R. and Low, D. R. (2003) "The Multicultural Muse: Culture, Creativity and Innovation," *International Journal of Cross Cultural Management*, 3 (2): 235–59.

White, S. (2002) "Rethinking Race, Rethinking Development," *Third World Quarterly*, 23 (3): 407–20.

Wiig, H. and Wood, M. (1997) "What Comprises a Regional Innovation System? Theoretical Base and Indicators," in J. Simmie (ed.), *Innovation, Networks and Learning Regions*, London: Jessica Kingsley, pp. 66–98.

Williams, M. (2005) "Democratic Communists: Party and Class in South Africa and Kerala, India," Ph.D. Dissertation, Department of Sociology, University of California, Berkeley.

Williams, R. (1960) *Culture and Society*, London: Chatto & Windus.

—— (1973) *Keywords*, Oxford: Oxford University Press.

—— (1977) *Marxism and Literature*, Oxford: Oxford University Press.

—— (1993) "The Idea of Culture" in J. McIlroy and S. Westwood (eds) *Border Country*, Leicester: National Institute of Adult Continuing Education.

Wilson, M. (1979) "Strangers in Africa: Reflections on Nyakyusa, Nguni, and Sotho Evidence," in W. A. Shack and E. P. Skinner (eds) *Strangers in African Societies*, Berkeley, CA: University of California Press, pp. 51–66.

Wilson, P. (2003) "Ethnographic Museums and Cultural Commodification: Indigenous Organizations, NGOs and Culture as a Resource in Amazonian Ecuador," *Latin American Perspectives* issue 128, 30 (1): 162–80.

Wolfe, D. A. and Gertler, M. S. (2001) "Innovation and Social Learning: An Introduction," in M. S. Gertler and D. A. Wolfe (eds), *Innovation and Social*

Learning: Institutional Adaptation in an Era of Technological Change, Basingstoke: Palgrave, pp. 1–24.

Woolcock, M. (1998) "Social Capital and Economic Development: Towards a Theoretical Synthesis and Policy Framework," *Theory and Society*, 27 (2): 151–208.

—— (2000) "Social Capital: Implications for Development Theory, Research, and Policy," *World Bank Research Observer*, 15 (2): 1–2.

Wordsworth, A. (2000) "Millions of Immigrants Needed to Sustain Economies," *National Post* (Canada) (March 22).

World Bank (1997) *Social Capital – the Missing Link*, Washington, DC: World Bank.

—— (1998a) *The Initiative on Defining, Monitoring and Measuring Social Capital*, Social Capital Initiative Working Paper 1. Washington, DC: World Bank.

—— (1998b) *World Development Report (1998/9): Knowledge for Development*, Washington, DC: World Bank.

—— (2004) "World Bank approves $34 million for Indigenous and Afro-Descent peoples: Development in Ecuador," World Bank Press Release no. 2004/426/LAC, June 17, 2004, accessed via www.worldbank.org on April 26, 2005.

World Business Council for Sustainable Development, Corporate Social Responsibility, 2005, accessed via http://www.wbcsd.ch/templates/TemplateWBCSD1/layout.asp?type=p&Menuld=Mz13&doOpen=1&ClickMenu=LeftMenu.

Worsley, P. (1999) "Culture and Development Theory," in T. Skelton and T. Allen (eds) *Culture and Global Change*, London: Routledge, pp. 30–42.

Xenos, Y. (1989) *Scarcity and Modernity*, London: Routledge.

Yakubu, U. (2002) "Cultural Erosion and the Crisis of Development in Nigeria," *Journal of Cultural Studies*, Nigeria, 4 (1): 1–55.

Yoshino, K. (1997) "Discourse on Blood and Racial Identity in Contemporary Japan," in F. Dikötter (ed.) *The Construction of Racial Identities in China and Japan: Historical and Contemporary Perspectives*, Honolulu, HI: University of Hawai'i Press, pp. 199–211.

Young, I. (1990) *Justice and the Politics of Difference*, Princeton, NJ: Princeton University Press

Young, L.A. (1996) "The Religious Landscape," in T. B. Heaton, T. A. Hirschl, and B. A. Chadwick (eds) *Utah in the 1990s: A Demographic Perspective*, Salt Lake City, UT: Signature Books, pp. 155–66.

Young, R. (1995) *Colonial Desire*, London: Routledge.

Index

adjustment policies/programs 51, 153, 166

Afghanistan 45

Africa: armed conflicts 60–1; cultural heritage 4; development projects 18, 58–9; *see also* Congo; Ethiopia; Ghana; Mozambique; Nigeria; South Africa

Allen, J. 54

Amazwi Abesifazane memory cloths project (KwaZulu-Natal) 204, 207–25; culture and development 221–3; funding and profits 209; as national memory archive 210–16, 218, 219–20; project aims 209–10; and social justice 216–21; use of traditional skills and culture 207–8

Amponsem, G. 132, 134–5, 136

Anderson, Benedict 44

Anderson, Perry 31

Andes/Andean countries: cultural economies 84–5; indigenous development 26, 229, 230–1; peasant systems 46, *see also* Bolivia; Bolivian musicians; Ecuador

anti-development initiatives 44

Appadurai, Arjun 114, 223

Arce, A. 11

Asia: *see* China; East Asia; India; Japan; Korea; Singapore; Southeast Asian "tiger" economies

Asmal, K. 210

Attacking Poverty (World Bank report 2000) 34

Australia 131, 132

Bandung Conference (1955) 38

Bassi, M. 76

Beck, Ulrich 142

Bennett, J. 224–5

Berlant, Laurent 45

Berman, Marshall 46

Besteman, C. 73–4

Bhavnani, J. 218

bilateral investment treaties 51

Boku Tache 71

Bolivia 121–2; crafts/knowledge 96; development/projects 85, 94–6; indigeneity 108, 122; indigenous peoples 86–7; international funding 21, 122; music economy 108, 110–11; natural resources 121–2; political movements/protests 9

Bolivian musicians: and "Andean" music 108–9, 110–13, 119; Japanese cultural differences 118; and Japanese tours 15, 27, 107, 108–19, 232; tour contracts/conditions 114–16, 124, *see also* Laughing Cats (*Warao Neko*)

Borana (Southern Ethiopia): CBNRM projects 69–72, 74, 77, 78–81; conflicts 73–4; failed projects 69, 72, 77; geography and borders 75–8; and government distrust 76–7, 78, 80–1; natural resources 73, 74; Peasant Associations 75, 77; refugees and returnees 76, 79; region and peoples 67–9, 73–6;

traditional institutions/leaders 70–2, 78–81, 230; UNHCR actions 76
Botha, Andries 206, 208, 209
Bourdieu, Pierre 36, 41, 129
Brazil 38, 113, 236
Brazzaville 45
Brink, Andrew 218
Britain 14, 41, 55, 131, 147
Buck-Morss, S. 54
Burke, Edmund 40

Canada 126, 127, 131
capitalism 13–14, 51; and community 45; industrial 40; market 52; modern 45–7, 170; post-Fordist 98–9, 114, 232; and profitability 46
CBNRM (community-based natural resource management): Borana (Southern Ethiopia) project 67–72, 74, 77–8, 79–81; and culture as resource 81; and decentralizing states' culture 62–4; post-conflict projects 60–2
Chari, Sharad 50
China/Chinese diaspora 141–2
Chirwa, W. 219
Church of Jesus Christ of Latter-day Saints (LDS Church) 176; organizational system 188–90; social ideals and teachings 183, 194–5, see also Mormons/Mormonism
citizenship 156; corporate 32; of marginalized groups 206–7; rights 94; of women 220, 223, 224
civil society 36, 44, 51, 57, 235; and social capital 130, see also global civil society
Cleaver, Frances 163
clusters/clustering: advantages 174; policies 196–9; regional industrial 170–2; spatial limited of cultural economy 195–9; successful regions 196; and transaction cost reductions 172–4
Coleman, J.S. 126, 129
collective identity 46
Colombia 90
colonialism 3, 11, 74, 85, 140; development 38; in South Africa 206, 221
commerce/commercialization 126, 140; and ethnic economies 84, 138
commodification 52
communitarian theory 44
communitas 46–7

community/ies: and culture 42–8, 229; and development initiatives 33, 155, 163; kin-based groups 163; in Mozambique 62, 65–6; and participatory development projects 59, 62–4; and politics 45; and power relations 152; self-governing 47; and traditional institutions 64–7, 86
conflicts 6, 7, 53; in Africa 60–1, 75
Congo 45, 61
corporate conduct and practice 33–4
corporate social responsibility (CSR) 32–4
Cosgrove, Dennis 37
Cowen, M. 37
Crang, P. 39
Create Africa South 209
cross-cultural enterprises 135–43; ethno marketing 138; examples 136; implications 143–5; and social capital 144–5, 148; workforces 137
Cuba/Cubans 131
cultural development: concept 18; festivals 96, 97; workshops 86, 97, 100, 159–60, 207, 213, 216, see also Amazwi Abesifazane memory cloths project (KwaZulu-Natal)
cultural differences/diversity 146, 235, 236, 237; and economic growth 93, 203; and power 122–3; and social capital 126, 127–8, 144; and transnational politics 119–23; and World Bank 150
cultural economy 3, 107–8, 237; concept 16, 18; and entrepreneurs 109–10, 122; indigenous 103–4, 234; interethnic 27, 107, 109, 231; networks 86–7, 142–3, 174, 182–7, 231; spatial limits of 172, 195–9; variations in 168, see also clusters/clustering; high-tech industries
cultural politics 39, 42, 45, 51, 54, 72–9, 108
cultural revolution 51–4
culturalization 39
culture: as community 42–8; concepts, definitions, theories 17–19, 22–4, 49, 151–2, 175, 203, 233–4, 236; as creativity 104, 205, 219, 224, 236, 236–7; critical accounts 12–13; debates 8; in development 16, 19–24, 39–42, 151–3, 233–7; economic and social justice relation 30, 204–5; as institution 233–7; reification 143, 205; as resource 81, 233, 236; social capital link 151–3

cultures: encounters between 108–9; geography 19–22, 37, 84; identity 86–7, 91, 97, 230; stereotypes 127–8, 222–3, 232

Davis, Mike 57
de Araujo, Manuel 58
Dean, Mitchell 55
Decade for Cultural Development *see* United Nations Decade for Cultural Development (1988–97)
democracy: and social capital 129, 130, 145–6; in South Africa 224; and UNDP 34–5
Derrida, Jacques 32
development: agencies 2, 20–1, 21, 91; classifications and indices 34; concepts/models 2–3, 6, 7, 13, 23; cultural turn/relationship 38–42, 52, 204, 222, 225, 228–37; debates/discourses 8, 12–13, 51; economics 13–16, 134–5, 144, 173; institutions 50, 59; intervention policies 3, 22; as modernity 39; postcolonial 44–5; as social imaginary 48–50; theory and practice 10–12, 25, 38–9, 48, 203, 204, 229
development projects: aid flows/programs 3, 62, 93; community-based natural resource management (CBNRM) 60–2; and culture/cooperation 164–7, 168, 223–4; design and implementation 23; donor interventions 167; group-based 154–69; participatory 58, 61, 63, 236; and social capital 151–3; transnational pro-indigenous 111
diasporas: and immigrant enterprises 141–2
Discipline and Punish (Foucault) 41
Duvenage, P. 219

Eagleton, Terry 39–40
Earth Summit (Rio de Janeiro 1992) 33
East Africa 18
East Asia 27, 51, 228, *see also* Southeast Asian "tiger" economies
economic dynamism 126–7
economic geography 13–16
economic growth: contributors 6; and cultural diversity 93; and industrial clustering 196; and social capital 129, 130; strategies 173–4, *see also* regional cultural structures
economic relations: and culture-as-creativity 236; and embeddedness 173; and globalization 98; and indigenous peoples 83, 102; and social capital 126

economy/economics: capital 129–30; and culture 13–16, 39, 51; Five Year Plans 23; institutional 235; Keynsian doctrines 31; laws of 31; and regulation 50, 51; theory 50; transaction costs 172–4, *see also* cultural economy; political economy
Ecuador 4, 18; Agrarian Development Law 85; agricultural census findings 88; development projects 85; natural resource rights 90; political movements 89, 90; pro-indigenous development (PRODEPINE) 94, 96–9, 229
education 6, 194, 217; of immigrant groups 147; and indigenous peoples 87, 88, 89, 100, 209; intercultural 87, 94, 99; training and technical assistance 97–8
"El condor pasa" (Robles/Simon and Garfunkel) 111–12
Ellmeier, Andrea 109
Elyachar, J. 30
employment 5; of Bolivian musicians 109, 114, 116, 118–19, 120; creation 126–7; cross-cultural 137; in high-tech industries 176–7, 191, 197; and immigrant enterprise 126; of indigenous peoples 100; post-Fordist 98
empowerment 44, 62, 73; black economic 218; development/rehabilitation projects 221, 225; of indigenous peoples 84; and social capital 151; of women 205, 206, 210
Enlightenment thinking 40, 49
entrepreneurship 102, 154, 217; cultural 15, 27, 96, 109, 122, 232; migrant 127, 134, 138, 141, 142, 148; social 92, 93, 95
environmental issues 26, 34; developmental degradation 58, 69–70, 79; green governmentality 50; and indigenous knowledge 61, 65; management 65, 67; policies/projects 58, 59, 61, 71, 91, 156, 163; responsibility 33; sustainable development 90
Ethiopia: borders and conflicts 74–5; decentralization policies 62–4; and development organizations 79–80; natural resource management projects 25–6, 60; Peasant Associations 75, 77; post-conflict development programs 61, 62, 229, *see also* Borana (Southern Ethiopia)
Ethiopian Orthodox Christian Church 63

ethnic economies 130–5, 231–2; concept 26, 143–4; discourse 135, 136; embeddedness 134–5, 144; and immigrant economy 127; and labor 136–7; and rainbow economies 144

ethnicity 3, 10, 21, 76, 128, 134; definition/foundations 74, 131, 133–4, 222; nationalist 59, 63, 144; reification 127, 143, 205; and resources 74, 80

Eurocentrism 5, 40

European Union (EU) 15

Evans, Peter 36

exclusion 12, 228, 237; and citizenship 206; and cultural economies 182; ethnic/racist 94, 103, 132, 236; from development 87, 93; political 87; and poverty 164; social 34, 93; and social capital 23, 153

Feith, Douglas 45

feminist movements: and development thinking 3, 8; postcolonial 12–13, 228; and social justice 206; Third World 5, 12–13

Ferguson, James 35, 50

Fine, Ben 152

Foucault, Michel 41, 50, 55

France, Tunisian immigrants 127

Fraser, Nancy 45, 206

free trade 51, 57

functionalism 129–30

Gaonkar, D. 49

Geertz, Clifford 41

gender 101

geographies: and development cooperation 164–7, 168; developmental 84, 167–9; imaginative 84–5, 99–100, 103, 153, 167; industrial clustering 145

Germany 127, 131, 138

Gerschenkron, Alexander 38

Ghana/Ghanaians: cultural harmony 154, 155, 231; development initiatives 26–7, 151, 153–67, 229; enterprises 134–5, 136; ethnic groups 132; farmer groups 160, 161, 165; funds, loans and credit 158–9, 160–2, 165, 168; group failures 166, 168; independence and government 153; Intermediate Means of Transport (IMT) 157; leadership/organization 161; Local Development Groups (LDGs) 154–62; migrations/mobility 166; NGOs (local/international) 155–7,

159–60, 165, 168; poverty issues 153, 156, 168; regions 155–7, 166; resources 156; transport equipment purchasing 157–60; Village Infrastructure Project (VIP) 157–60; women's trader associations/groups 159, 161, 165; World Bank/IMF programs 153, 154–5, 157, 166

Giroux, H. 51

Glacken, Clarence 30, 31

global civil society 53

Global Positioning Systems (GPS) 65

global South: and communities 45; cultural development policies 26, 27; and governance 56; loss of cultural diversity 4

globalization 57; and economic relations 98; and homogenization 114; of labor 123–4; and livelihoods 59; threat to cultural diversity 92

governance 6, 34, 36, 92, 95, 235; and cultural development 22–3, 56, 236, 237; good 103, 130, 145, 231, 235; neo-liberal 229, 235; and social capital 145

governmentality: colonial 203; green 50; modern 55

Grameen bank (Bangladesh) 155

Granovetter, M. 14–15, 20

Graybill, L. 220

Guatamala, Manos Campesinas (farmers' association) 93–4

Gupta, A. 48

Hannerz, Ulf 121

Harriss, L. 35–6, 151

Harvey, David 52

Herder, Johann 40

high-tech industries 170–2; cluster policies 196–9; computer software 176; and cultural economy 171–2, 174, 176, 180, 187, 190, 191, 199; inter-firm cooperation 173–4, 180–2, 183, 198–9; and spatial limits of cultural economy 195–9; Utah case study 176–95

Hishimoto, Koji 110, 112, 115, 116

Human Development/Poverty Indices 34

human rights: abuse and violation 203, 224; cultural and religious 62; of indigenous peoples 87, 90, 95–6; movements 53; and UNDP 34–5

identity/ies: collective 46; cultural 86–7, 89,
91, 97, 99, 230; modern 59; national 45
Ignatieff, M. 219
Image World of Consumer Capitalism
movement 53
Imagined Communities (Anderson) 44
immigrant enterprises 26, 135–6, 140–5; and
bank credit 138; cross-cultural dimension
139–41, 147; diaspora chains/networks
142–3; and domestic enterprises 126–7;
implications 143–5; long-term outlook
139–43; policy implications 145–8; and
social capital 144–5; stereotypes 127–8;
typology 138–9; understanding 146–7,
see also cross-cultural enterprises; ethnic
economies
immigrants/immigration: and economic
dynamism 126–7, 138; and ethnicity 134;
illegal/human smuggling 127, 142;
integration 141–2, 147; laws 127
India/Indians 49, 140, 142, 147
indigeneity 53, 108
indigenous development 26, 58–9, 83–5,
98–104; changing paradigms 91–4; funding
88–9, 101–2; imaginative geographies 84–5,
99, 100, 103; marginalization and inclusion
94–9; socio-spatial fixing 99–103; and
women's issues 100–1
indigenous peoples: cultural identities 86–7,
89, 99; empowerment 84; and human rights
87, 90; knowledge 61, 65; languages 64, 67,
87, 92, 95, 97, 101, 161; ownership rights
98; productive enterprises and networks
86–7; "traditional" music exports 94; urban-
dwelling 102–3
industrial clustering 145, 170–2, 232, *see also*
clusters/clustering; Utah's Wasatch Front
high-tech industry
institutions, in high-tech cluster policies
197–8
Inter-American Development Bank (IADB)
91; funding 93, 96; programs 4
Inter-American Foundation 236
international development organizations 59,
62
International Fund for Agricultural
Development (IFAD) 97
International Labor Organization (ILO) 90
International Monetary Fund (IMF) 31, 35
The Interpretation of Cultures (Geertz) 41

Ireland 131
Irwin, B. 71
Islam 9, 28, 45, 46, 133, 134; diaspora 141–2;
political/militant 47, 54, 57
Israel 132
Italy 127
Ivory Coast 61
Iyer, Pico 142–3

Jackson, Peter 37
Japan 51; "Andean" (folklore) music 108, 109,
112, 113; and Bolivian music tours 107,
119–24; cultural isolation and "return to
Asia" 119–21; foreign workers'
classifications/visas 114, 119–20, 121, 124;
music genres and audiences 111, 112–13,
119; nationalism and homogeneity myth
119, 123, *see also* Laughing Cats (*Warao
Neko*)
Jews 126, 132, 140
The Jews in Modern Capitalism (Sombart)
126
Joseph, M. 45
justice: and injustice 206; recognitive 45,
see also social justice

Keane, J. 47, 53
Kliksberg, B. 2
Korea 120, 132–3
KwaZulu-Natal 209, 210–12; factional
violence 210–12; gendered violence 213–14,
216–17; HIV/AIDS devastation/ostracism
214, 216, *see also* Amazwi Abesifazane
memory cloths project

labor: and ethnic economies 136–7; gendered
divisions 84, 101; globalization 123–4;
Japanese migrant work classifications 114;
in OECD countries 127; post-Fordist
employment 98–9, 114; relations 33;
undervaluation of indigenous peoples 88;
working hours 185–7, *see also* employment
Laclau, Ernesto 47
land: community rights 66, 97; tenure
systems/security 62, 65, 67, 70, 86;
traditional management 70
Lash, S. 31
Latin America: citizenship rights 94; cultural
economies 84–7, 90, 95; cultural identities
91; development projects 91–4; health and

education 88, 100–1; and human rights 90, 95–6; indigenous peoples 26, 83–5, 91–103; international aid 91; "Lost Decade" 89; political movements/protests 89–90; poverty and employment 87–8, 100–1; rural-urban migration 20, 89; social capital resources 93, 101; *see also* Andes; Bolivia; Colombia; Cuba, Equador; Guatamala; Peru

Laughing Cats (*Warao Neko*) 110, 114–19, 123–4; labor conditions/contracts 114–15, 124; tour groups/routines 115–18, 123

leaders/leadership 49; and authority 188–9, 190; indigenous 1, 70, 72, 74, 102; styles 6, 89, 92, 161; traditional 58, 62, 65, 66, 67, 70, 72, 78, 80, 103; women 161, 163

Lewis, Simon 217–18

Li, Tania 47, 50

Local Development Groups (LDGs) 154–7

local exchange trading system (LETS) 147

Long, Norman 11

Los Angeles 137

Los Angeles Times 137

Ludden, D. 49

Machel, Samora 59

McKay, David O. 184

McNeill, Desmond 130

Malthus, Thomas 54–6, 229

marginalization: exclusion/inclusion 151; of indigenous peoples 86, 87–8; and pro-indigenous development 94–9; of women 206

markets/marketing: Chartism/Owenism 52; |and commodities 52; and cultural economy 107–8; international trading 99; place marketing 84

Marxism 3, 84, 173; economy 41, 91

Masden, Kirk 119

Mayer, M. 154–5

Mbutu 38

memory cloths *see Amazwi Abesifazane* memory cloths project (KwaZulu-Natal)

Metahistory (White) 41

migrants/migration: asylum seekers 147–8; and capital link 146–7; displaced persons 60; environmentally induced 61; Japanese migrant work classifications 114; rural-urban 20

Mitchell, Timothy 50

modernity 37, 48, 236; alternative 50; capitalist 47; and development 39, 43, 203; and ethnic economy 135; Western 49

modernization 3, 10–11, 89, 235; models 6; and modernity 46; theory 38, 44, 203

moral economy 46, 55

Morita, Akio 143

Mormons/Mormonism 171; church/family/work priorities 184–7, 192–3; culture/traits 22, 27–8, 177, 187–90; economic performance 191–3; impact on regional economy 179–80; regional industry case study 176–95, 232; self-sufficiency trait 182–7; top-down leadership 187–90; unity/trust ethics 180–2, 189; work patterns 184–7, *see also* Utah's Wasatch Front high-tech industry

movements 53–4; anti-globalization 54; cooperative 52; feminist 3, 5, 8, 12–13, 206, 228; oppositional 57; political 9, 89, 90; women's 34, 53, 207

Mozambique: community-based projects 65; natural resource management 60; post-conflict development programs 61, 62; traditional institutions 64

multinational corporations 171; and insiderism 143

Music of the Masters (*Música de Maestros*) 108, 110–11, 112–13

Muslims 133–4, 138, 142

Myrdal, Gunnar 11

nation-states 10, 89, 124, 173; creation 45; and culture 5, 16, 20, 96, 109, 231

National Conference on Culture (Maputo 1993) 67

national identity 45

nationalism 40, 44–5, 52, 123, 124; and conflict 81

natural resource management (NRM) 60–2, 230, 235; Ethiopian projects 25–6, 60; in Mozambique 64, *see also* CBNRM (community-based natural resource management) projects

natural resources *see* resources

neo-liberalism 31–2, 85, 89, 229; and culture 51–3, 54–7, 90–1, 104; decentralization 62–4; deregulation 51, 92; development 23, 56, 92–3, 98, 205, 221, 234; global 31, 48, 52, 53, 56–7; model, theory and origins 34,

44, 47, 51–2; policies/programs 13, 62, 100, 103, 228–9, 235; political economies/reforms 22–3, 95, 122, 124, 232
Nigeria 46, 47, 153
Nixon, Richard 51
non-governmental organizations (NGOs) 20, 34, 44, 69; and cultural insensitivies 162; and development projects 59; and global civil society 53
"Notes on the Global Ecumene" (Hannerz) 121
NRM see natural resource management
Nyerere, Julius Kambarage 38

Ong, Aihwa 109, 114
The Origins of Our Time (Polanyi) 52
Oromo Liberation Front (OLF) 77
Outline of a Theory of Practice (Bourdieu) 41

Paine, Thomas 54, 55–6
Parsons, Talcott 10, 135
participation 2, 18, 73, 155, 224; of civil society 21, 235; community 43, 60; ethnic/indigenous 87, 90, 96, 132; as key concept in development 22–4; paradigms 25; and social capital 130; theories 61–2
participatory development projects 58, 61–2, 63, 71, 84, see also CBNRM (community-based natural resource management) projects
Partridge, W. 92
Peru 21, 87, 112, 236
Polanyi, Karl 52, 135, 173
political economy 7, 54, 231; capitalist 84, 229; and development 17; law 26; neo-liberal 22–3; post-Fordist 114, 232
politics 43, 53; anti- 37, 56, 57; generative 48, 57; international 28, 74; transnational 119–23, see also cultural politics
Porte Alegre (Brazil) 37, 236
Porter, Michael 196, 198
Portes, A. 145, 146
post-Cold War 41, 53, 60; Berlin Wall collapse 31
post-Fordist capitalism 98–9, 114, 232
post-structuralism, development discourse analysis 12
post-Washington consensus 36, 229
postcolonialism: anti-discrimination measures 5; indigenized development 38;

legacies of cultural interaction 2; theory 25, 48–9
poverty: eradification/reduction 37, 67, 153; of indigenous peoples 86, 87–8, 92; and migration 147; politicized 54; poor laws 54, 55; and scarcity 55–6; and social interaction 151
power 54; in communities 152; and cultural differences 122–3; and trust 161–2
Prabhu, A. 222, 223
Prague Spring 53
privatization 51
PRODEPINE I/II (Development Project of Indigenous Peoples and Afro-Ecuadorians of Ecuador) 96–103
Putnam, Robert 52, 126, 145; and social capital 128–9, 151, 152

radicalism 32, 40, 44; popular 54
Rankin, K.N. 151, 154–5
Rao, V. 204, 225
refugees 79, 140; from armed conflict 60–1; returnees resettlement 60–1, 76
regional culture structures: clustering 170–2; economic performances 199–200; interaction networks 174, 182–7; scalar hierarchy 175–6, 199; transaction costs reduction 172–4, see also Utah's Wasatch Front high-tech industry
religion: and culture 9, 12, 131, 133, 195; and ethnicity 133, 137; and human rights 62; and labor discrimination 194–5
resources: community-based management 44; and conflicts 60; ethnic rights 91; management 25–6, 60–2, 64; marketing/trading 6–7, 88, 99; natural 60–2, 73, 90–1, 121–2; and social capital 93, 146
Rights of Man (Paine) 54
Robles, Daniel Alomía 111
Rosaldo, Renato 121
Rose, N. 43
Rostow, Walt 10, 38
Rumsfeld, Donald 45
rural development programs 50

Said, Edward 48, 111
Salt Lake City (Utah) 176
Sánchez de Lozada, Gonzalo 121–2
sanctions 27, 159, 164–5, 167; and trust 162, 163, 166, 168

Sassen, Saskia 114
Save the Children US 70
Saxenian, AnnaLee 173, 174, 184
Scandinavia 131
scarcity 55–7, 229
Schlee, G. 73
Sen, Amartya 19, 34, 38
Sewell, William 39, 40–1
Shenton, R. 37
Silicon Valley (California) 14, 174–5, 184, 196; and copycat clusters 170–2
Singapore, development 6
social capital 128–30; causes and outcomes 127–8; concepts 23–4, 130, 150, 168–9, 231; and culture linkage 144–5, 151–3; and development 151–3, 168; and economic prosperity 129; and good governance 145; of indigenous peoples 101; as key resource 93; models and studies 23, 152, 231; and resources 146; and World Bank 35–7; World Bank's interpretation 27
social imaginary 42, 43, 44; development as 48–50
social justice 204, 205–7, 221, 225; *Amazwi Abesifazane* memory cloths program 205–25
social relations: cross-border/transnational 142–3; cultural implications 26; and social capital 151
socialism 38, 47, 63
socioeconomics 9, 137; development 85, 233, 237; dynamics of change 11, 15, 24; inequalities 121; levels/status 137, 144
Somalia/Somali people 45, 73–4
SOS Sahel (UK NGO) 70, 71, 79
South Africa: apartheid memories/legacies 28, 206–7, 210–12, 220–1, 232–3; citizenship 206–7; cultural stereotypes 222–3; developmental and rehabilitation programs 207, 217, 221–3, 225; gender inequality and injustice 224–5; human rights violations 207, 224; National Women's Coalition 207; post-apartheid development/reconstruction 28, 216–17, 223; *sangoma* (traditional healers) 28; Self-Employed Women's Union (SEWU) 207; Truth and Reconciliation Commission (TRC) 207, 211, 217–18, 220–1, *see also Amazwi Abesifazane* memory cloths project (KwaZulu-Natal)
Southeast Asian "tiger" economies 203, 228; as development models 6, 7

Stanley, E. 220
The Stranger (Simmel) 126
sub-Saharan Africa 20, 155
Sudan 61
Summers, Lawrence 31
SwissAid 88

Taylor, Charles 25, 49
Third World: feminist movements 5, 12–13; and neo-liberalism 52
Toronto (Canada) 137
tourism 18, 122, 222; eco-/ethno- 84, 94, 96, 103, 138, 234; handicraft-based 96, 97, 231
trade/trading: cross-cultural 126, 139–40; diasporas 142–3; LETS system 147, *see also* markets/marketing
tradition 58–60; and community 43; cultural 12, 38; and modernity 24, 203, 235, 236
traditional institutions 58, 59–60, 80; in Borana region (Ethiopia) 69, 70–1; and communities 64–7; development approach 72–3; in development thinking 235
transnational corporations (TNCs) 32–4
Truman, Harry S. 37–8
trust: and cooperation 161, 162–3, 168; and cultural unity 180–2; and donor insensitivity 162–5; power and control processes 161–2; and social capital 168
Tsing, Anna 50

UN Codes on Conduct for Law Enforcement 33
UN International Covenants on Civil and Political Rights and on Economic, Social and Cultural Rights 33
UN Universal Declaration of Human Rights 33
UN World Investment Report 51
United Nations 4, 5, 7; and human rights 34–5, 90
United Nations Decade for Cultural Development (1988–97) 2, 4, 18, 229
United Nations Development Program (UNDP) 4, 32, 34–5; *Human Development Report* 34, 35; joint World Bank programs 35
United Nations Food and Agriculture Organization (FAO) 65
United Nations High Commissioner for Refugees (UNHCR) 76

United States (USA) 126, 131, 133, 134;
American hegemony 32, 53; and forced
nation building 45; National Security
Strategy (2002) 52; and neo-liberalism 51–2
Uquillas, J. 92
Urry, J. 31
US Agency for International Development
(USAID) 30, 31
US Civil Rights Act (1964) 194
US National Center for the Educational
Quality of the Workforce (EQW) 187
US Religious Freedom Restoration Act (1997)
195
US Workplace Religious Freedom Act (1972)
195
Utah's Wasatch Front high-tech industry
176–95, 199; economic performance 191–3;
employee involvement 188; family vs work
commitments 184–7, 192–3; innovation and
creative dissent 187–90; labor discrimination
194; regional culture/mechanisms 177–8,
193–5, 199–200; strategic partnering 182–4;
subsectors 176, 177; survey dataset 178–9;
workforce 177, 186–7, see also
Mormons/Mormonism

Walton, Michael 204, 225
Washington consensus 13, see also post-
Washington consensus
Watts, Michael 203
Weber, Max 65
welfare 15, 51, 151
Western Europe 51, 172
White, Hayden 41
Williams, Raymond 39, 40, 43–4, 57, 222
Wolfenson, James 21, 51

Wolfowitz, Paul 38, 45
women: black issues 69, 206–7, 211, 212,
217, 218, 220–1; citizenship 220, 223,
224; crafts and neighbourhood groups
204–5; and development projects 13, 97,
100–2, 223; discrimination/marginalization
69, 206, 224; education and employment
12, 100–1; empowerment 205, 206, 210;
gendered divisions 13, 84, 101–2, 163,
224; HIV/AIDS devastation/ostracism
214, 216; and indiginous development
100–1; inequality issues 12, 69, 163;
movements 34, 53, 207; resource
marketing/trading 86–7, 88, 99; and social
justice 205–10; violence/ abuse against 209,
213–14, 216, 224, 233; voices of 28,
216–17, 220–1, 223; workshop processes
207, 216, see also Amazwi Abesifazane
memory cloths project (KwaZulu-Natal)
World Bank 69; culture-development
integration 67; and development programs
50, 92, 150; developmental classifications
34; "greening" 50; and indigenous
development 91, 93, 102; interpretation of
social capital 27; Latin American loans 21;
and social capital 35–7, 129, 130, 145, 150,
154, 155, 167; social theory 31, 32;
Structural Adjustment Policies (Ghana) 153,
166; and UNDP 35; World Development
Reports 34
World Business Council for Sustainable
Development 32–3

Xenos, Y. 56

Young, I. 206